T0200601

Pumps for Chemical Processing

J. T. McGUIRE

Dresser Industries, Inc.
Huntington Park, California

CRC Press
Taylor & Francis Group
Boca Raton London New York

CRC Press is an imprint of the
Taylor & Francis Group, an **informa** business

CRC Press
Taylor & Francis Group
6000 Broken Sound Parkway NW, Suite 300
Boca Raton, FL 33487-2742

First issued in paperback 2019

© 1990 by Taylor & Francis Group, LLC
CRC Press is an imprint of Taylor & Francis Group, an Informa business

No claim to original U.S. Government works

ISBN-13: 978-0-8247-8324-2 (hbk)
ISBN-13: 978-0-367-40314-0 (pbk)

This book contains information obtained from authentic and highly regarded sources. Reasonable efforts have been made to publish reliable data and information, but the author and publisher cannot assume responsibility for the validity of all materials or the consequences of their use. The authors and publishers have attempted to trace the copyright holders of all material reproduced in this publication and apologize to copyright holders if permission to publish in this form has not been obtained. If any copyright material has not been acknowledged please write and let us know so we may rectify in any future reprint.

Except as permitted under U.S. Copyright Law, no part of this book may be reprinted, reproduced, transmitted, or utilized in any form by any electronic, mechanical, or other means, now known or hereafter invented, including photocopying, microfilming, and recording, or in any information storage or retrieval system, without written permission from the publishers.

For permission to photocopy or use material electronically from this work, please access www.copyright.com (http://www.copyright.com/) or contact the Copyright Clearance Center, Inc. (CCC), 222 Rosewood Drive, Danvers, MA 01923, 978-750-8400. CCC is a not-for-profit organization that provides licenses and registration for a variety of users. For organizations that have been granted a photocopy license by the CCC, a separate system of payment has been arranged.

Trademark Notice: Product or corporate names may be trademarks or registered trademarks, and are used only for identification and explanation without intent to infringe.

Library of Congress Cataloging-in-Publication Data

McGuire, J. T.
    Pumps for chemical processing / J. T. McGuire.
        p.   cm.
    Includes bibliographical references and index.
    ISBN 0-8247-8324-7 (alk. paper)
    1. Pumping machinery.   I. Title.
TP156.P8M34   1990
660'.283--dc20   90-41553
                                          CIP

Visit the Taylor & Francis Web site at
http://www.taylorandfrancis.com

and the CRC Press Web site at
http://www.crcpress.com

Pumps are among the oldest and most widely used of the devices employed by mankind in its efforts to raise its standard of living. Not surprisingly then, pumps have been the subject of a great many books. Most of these books have dealt with a particular pump type, and in many instances with a particular aspect of one pump type. When a publication has dealt with pumps generally, it has tended, out of a desire for completeness, to be a tome.

Chemical processing uses all three basic pump types, namely centrifugal, rotary, and reciprocating. That warrants a comprehensive text with a concise treatment of each pump type. To be a useful single reference for the chemical engineer, it has to cover application, selection, construction, procurement, installation, operation, and maintenance.

To gain the required coverage, this book has been prepared as a guide to determining a pump's duty; selecting the most appropriate pump type, materials, and seal; choosing the manufacturer; and installing, operating, and maintaining the pump. Throughout, emphasis has been placed on those aspects that experience shows to be the cause of practical operating problems. Although *Pumps for Chemical Processing* focuses on chemical engineering, the necessary breadth of its coverage means that much is applicable to pumping in general.

As is always the case, the book's preparation was a cooperative effort. Dresser Industries, Inc., provided much of the impetus for it and contributed handsomely to the typing of the manuscript and the preparation of the illustrations. In this connection I am particularly indebted to Gus Agostinelli for both encouragement and assistance; Eileen Goldrick for typing the manuscript; Craig Williams, Fred Paine, and Jack Doolin for organizing many of the illustrations; and Igor Karassik and Will Smith for proofreading and critiquing in their areas of expertise.

For the composition of the manuscript, my wife, Karin, provided what I feel is most valuable to those trying to write: tolerance of remarkable swings in mood and long periods of neglect.

J. T. McGUIRE

Contents

Pumping in the chemical process industry involves the bulk movement of process liquids, the precise injection of reactants, and the provision and dissipation of energy. For convenience, these three functions are referred to as process, metering, and utility. While utility does not involve the handling of process chemicals, there will be no processing without it, so its inclusion is justified.

Pumping is the so-called "heart" of chemical processing, and it's a good analogy. Successful pumping is thus of fundamental importance; achieving it involves getting all the following "right":

1. Conditions of service
2. Pumping specification
3. Procurement
4. Installation
5. Operation
6. Maintenance

Not getting any one of these "right" can cripple a process. Generally, though, mistakes in the last four can be corrected, albeit at great expense. But if the first two are not "right," if the basics are not sound, all subsequent corrections will be "Band-Aids."

Because developing a sound pump specification is so important, it is the focus of this section. Figure 1.1 shows the basic sequence for doing this, which this volume follows. Note, also, that iteration is an inherent part of this sequence.

The pump types reviewed are centrifugal, rotary, and reciprocating. For each type the review covers basic characteristics, specific configurations and their uses, then available materials of construction, all with a view to chemical process service. For clarity, specific information on metering and utility pumps is given in Chapters 9 and 10, respectively.

The size of this volume is limited, thus an exhaustive treatment is not possible. Instead, the material seeks to highlight the fundamental, the novel, and the usually troublesome, thereby supplementing other related material. For greater detail on pumps in general, the reader is referred to the *Pump Handbook* [1.1], *The Chemical Engineering Guide to Pumps* [1.2], *Centrifugal Pumps* [1.3], and *Reciprocating Pumps* [1.4]. A recent publication, *Process Pump Selection—A Systems Approach* [1.5], adds notably to the "art" of pump selection (in metric units). Throughout the volume, references on specific topics are cited to enable further reading.

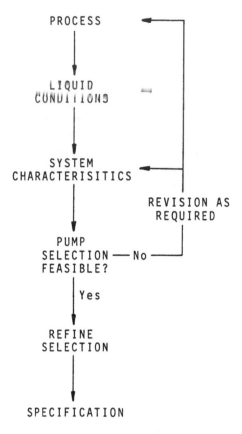

G. 1.1.   Process for developing a sound pump specification.

References

1.1.   I. J. Karassik et al. (eds.), *Pump Handbook*, 2nd ed., McGraw-Hill, New York, 1986.

1.2.   K. McLaughton (ed.), *The Chemical Engineering Guide to Pumps*, McGraw-Hill, New York, 1984.

1.3.   I. J. Karassik and R. Carter, *Centrifugal Pumps*, McGraw-Hill, New York, 1960.

1.4.   T. L. Henshaw, *Reciprocating Pumps*, Van Nostrand Reinhold, New York, 1987.

1.5.   J. Davidson (ed.), *Process Pump Selection—A Systems Approach*, The Institution of Mechanical Engineers, Suffolk, U.K., 1986.

2. Liquid Conditions

Chemical process pumping involves the handling of liquids that are corrosive or toxic or both. This requirement distinguishes the service from general pumping and has a major bearing on materials of construction, internal mechanical construction, and the type of seal necessary.

Because the nature of the liquid to be pumped has such a bearing on pump construction, its precise determination is an essential first step in any pumping application. Not doing this with sufficient precision is a prime cause of premature failure in chemical pumps.

2.1. Properties

Liquid properties influence the type of pump and its mechanical construction. Those needed for pump selection are:

Specific gravity (SG) or relative density (RD)
Vapor pressure
Viscosity
Rheological characteristic (if other than Newtonian)

Specific heat, though not often quoted, is useful, particularly when the application has minimal net positive section head (NPSH) available.

Liquid properties are usually specified at the pumping temperature or over the expected temperature range, if that is the case.

2.2. Temperature

Liquid properties and corrosiveness vary markedly with temperature, thus the precise temperature is important. General terms such as "cold," "hot," or "ambient" do not provide sufficient information. An ideal specification gives the expected temperature range and the normal operating temperature.

2.3. Constituents

Most liquids pumped are solutions of multiple constituents. To help insure the most appropriate materials are used for the pump, it is necessary to know the liquid's constituents and their concentration. In this connection it is vital that all constituents, major and trace, be identified and that their concentration be given in specific units.

Trace constituents, particularly halogens, halides, or compounds of hydrogen, can render a nominally suitable material entirely unsuitable. Examples are the effect of fluorides on high silicon iron (cited by Birk and Peacock in Ref. 1.1) and of hydrogen sulfide on a variety of materials (see NACE [2.1]).

Concentration needs to be stated specifically to avoid varying interpretations of terms such as "dilute" and "concentrated." Similar treatment is necessary for trace constituents because their effect can vary markedly with very small changes in concentration.

2.4. Acidity and Alkalinity

Whether a solution is acidic or alkaline or likely to vary is of consequence to material selection. For this reason, the solution's pH, or possible range of pH, needs to be specified.

2.5. Aeration

The degree of aeration of a solution can have a significant effect on its corrosiveness. Alloys that rely upon oxidation for passivity, Type 316 stainless steel for example, suffer severe corrosion in deaerated solutions. For alloys dependent upon a reducing environment for corrosion resistance, aeration of the solution can promote severe corrosion.

2.6. Solids

In low concentrations, often seemingly innocuous, solids suspended in the pumped liquid can cause corrosion-erosion. Frequently, the damage can be severe enough to result in premature, catastrophic failure of a pump's casing. If solids are likely to be present, their material, size, and concentration need to be specified.

2.7. Allowable Leakage

Pollution, both atmospheric and ground, known carcinogenic effects, and high toxicity dictate that many of the liquids used in the chemical industry be allowed to leak at only very low rates or not at all. Low or zero leakage requires special consideration in the selection, design, and quality of the pump's "pressure boundary."

2.8. Product Quality

With some liquids, their quality, either purity or condition, may be affected by the pump by way of contamination or agitation, respectively. When this is the case, it needs to be clearly specified so the appropriate pump configuration and materials can be chosen.

2.9. "Other" Characteristics

Some processes involve the pumping of liquids with special or abnormal characteristics not covered in the preceding material. An example is polymerized resins which will eventually "set" if allowed to stand in the pump. Avoiding difficulty with such an application imposes special requirements on pump selection, installation, and operation, all of which may be overlooked if the characteristic is not identified at the start of the project.

Reference

2.1. *Sulfide Stress Corrosion Cracking Resistant Metallic Material for Oil Field Equipment*, Publication MR-01-75, National Association of Corrosion Engineers, Houston, 1975.

3. Corrosion

A general appreciation of the nature of corrosion is important in selecting pump materials, reviewing various designs, and correcting field problems. Corrosion is very broad subject, so the treatment following is cursory in the extreme.

Corrosion in pumps differs from general practice in two important respects. First, liquid velocities within pumps are often higher than in

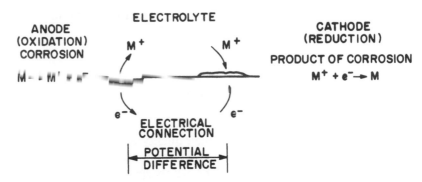

FIG. 3.1. Galvanic corrosion. Two sites with a potential difference, electrically connected, and in the presence of an electrolyte produce corrosion by oxidation at the annode, reduction at the cathode.

pipelines and vessels, so corrosion data based on low velocity tests may not be applicable to pump parts. Second, some pump components, e.g., seals and shafts, cannot tolerate any appreciable penetration or weight loss without failure, thus, they may have to be made of materials superior to those in the rest of the system.

Seabright and Fabian [3.1], drawing from Jastrzebski [3.2], broadly define corrosion as deterioration of materials by chemical or electrochemical action. Van Vlack [3.3] narrows the definition to "deterioration of solids by liquid electrolytes," or electrochemical action.

Materials (solids) used for pump construction fall into two categories: metallic and nonmetallic. Corrosion, by the narrower definition, is almost exclusively limited to metallic materials, whose electrons are free to move. Nonmetallic materials generally do not have such electron freedom, thus any deterioration is usually by chemical action alone.

3.1 Metallic Corrosion

While there are various special cases of metallic corrosion, all are fundamentally galvanic in nature. Referring to Fig. 3.1, the essential ingredients are a potential difference between two sites immersed in an electrolyte and connected with an external electrical circuit. The potential difference causes oxidation at the anode (metal ions go into solution and electrons move into the external circuit) and reduction at the cathode (electrons from the external circuit reduce ions from solution).

As in electroplating, the rate of anode consumption or corrosion depends upon current density. Current density, in turn, depends upon the potential difference and balanced oxidation/reduction reactions. These dependencies

TABLE 3.1 Corrosion Type Versus Basic Galvanic Cell

Corrosion Type	Galvanic Cell		
	Unlike Electrodes	Stress	Concentration
General, direct	Δ		
Crevice			Δ
Pitting			Δ
Stress cracking		Δ	
Corrosion fatigue		Δ	Δ
Intergranular	Δ		
Galvanic	Δ		
Erosion-corrosion	Δ		
Selective leaching	Δ		
Microbiologically induced			Δ

lead to means of limiting corrosion rates. First, the potential difference can be reduced. Second, the oxidation/reduction balance can be kept at a very low level. Two means are available to achieve the latter: cathodic polarization, in which the reduction rate limits the balance, and passivation, in which initial oxidation of the anode renders it essentially inactive.

A mechanical factor of great consequence to the oxidation/reduction balance is the relative size of anode and cathode; a relatively small anode is susceptible to rapid corrosion. In this connection, materials protected by passivation are vulnerable if the passive film is perforated.

Corrosion damage is generally identified by 10 types. While in some cases the whole mechanism is not yet fully understood, these 10 types of corrosion are the products of three basic galvanic cells. Table 3.1 categorizes the various corrosion types by their basic galvanic cell.

3.2. Galvanic Cells

Table 3.1 shows most corrosion is produced by the unlike electrode cell. Dissimilar materials, far apart on the galvanic series, in the presence of an electrolyte is the classical model of the galvanic cell, and produces what is generally known as galvanic corrosion. Unlike electrodes, however, do not have to be large or separate; they can be different phases within the same structure, e.g., ferrite and cementite in iron, see Fig. 3.2. A second example is regions of different composition, e.g., those deficient in chromium in sensitized austenitic stainless steel, see Fig. 3.3, which lack passivity and are thus susceptible to intergranular corrosion. To an even finer degree, potential differences caused by such subtleties as crystal orientation and the presence of precipitates are cited in explanation of general corrosion.

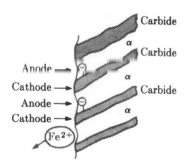

FIG. 3.2. Galvanic microcell (pearlite). Since the two phases are different in composition, they have different electrode potentials and produce a small galvanic cell. (Reprinted with permission from Ref. 3.3.)

Selective leaching occurs when metals whose structure is a matrix of dissimilar materials are exposed to an electrolyte.

Graphitization, in which the iron (ferrite and cementite) in grey cast iron becomes anodic to the flake graphite, is a form of selective leaching. An insidious aspect of graphitization is the tendency of the graphitized surface to gradually become noble (cathodic) to other materials in the pump. In this manner an "iron" casing can eventually cause the corrosion of a bronze impeller. High zinc brasses are most susceptible to selective leaching but are little used in chemical processing.

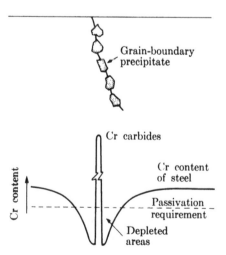

FIG. 3.3. Chromium depletion adjacent to the grain boundary. The carbide precipitation consumes nearly 10 times as much chromium as carbon. Since the larger chromium atoms diffuse slowly, the Cr content of the adjacent areas is lowered below protection levels. (Reprinted with permission from Ref. 3.3.)

Erosion-corrosion is general corrosion whose rate is accelerated by erosion. The effect of erosion is to continually remove the products of corrosion, thus eliminating any protective effect they may have had.

Regions of high energy or tensile strain are anodic to their surroundings, giving rise to the second type of galvanic cell, the stress cell. Anodic regions are typically grain boundaries, dislocation lines, imperfections, and sites of high residual or imposed stress. While the basic mechanism of *stress cracking* seems clear, the detailed mechanism is not, consequently the prediction of stress cracking still relies to a great extent on empirical data.

Corrosion fatigue refers to the phenomenon whereby a component fails by fatigue but at a stress level well below its atmospheric endurance strength. The mechanism of failure appears to be accelerated crack propagation by corrosive action (stress cell or concentration cell or both). Guy, cited by Ref. 3.1, notes corrosion fatigue "damage ratios" (corrosion endurance strength divided by atmospheric endurance strength) in the order of 0.5 for "stainless steel" in salt water. Kratzer [3.4] reports 0.7 for a 27 chrome, 0.2 carbon steel in polluted river water.

Corrosion fatigue and stress cracking are insidious; both can cause catastrophic failure at nominally safe stress levels in an ostensibly safe environment.

The third basic galvanic cell, the *concentration cell*, develops a potential by a difference in electrolyte concentration. A particular version of this cell, the *oxidation* type, is the one that usually afflicts pumps. Either by design or as a product of debris accumulation, some regions become oxygen deficient, see Fig. 3.4. With materials dependent upon passivity for corrosion resistance, the oxygen-deficient regions become anodic and susceptible to rapid corrosion. The concentration cell is thought to be a factor in *pitting corrosion*, serving to accelerate the rate of corrosion as the pitting deepens, see Fig. 3.4.

Microbiologically induced corrosion (MIC) refers to corrosion involving the action of bacteria on metal surfaces. It may be considered a form of concentration cell attack in that the microbe alters the bulk environment in some manner within the growing colony. A variety of aerobic and anaerobic bacteria have been associated with attack on individual alloy systems. The specific mechanism will vary. A problem of current technological importance is the severe pitting that has been reported in stainless steel vessels and piping, usually following prolonged exposure to stagnant water. It has been attributed to a microbe that oxidizes inorganic compounds and concentrates halides. Ferric and manganic chlorides are formed, and these induce pitting beneath the microbe colony.

3.3. Other Metal Deterioration

The types of corrosion covered so far are those clearly caused by electrochemical action. There are two further types of metal deterioration often associated with electrochemical action but not clearly caused by it.

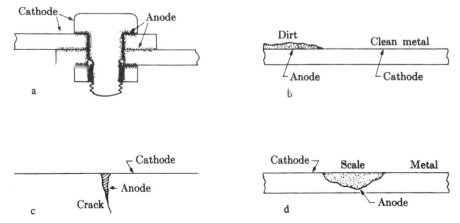

FIG. 3.4. Oxidation cells. Inaccessible locations with low oxygen concentrations become anodic. (Reprinted with permission from Ref. 3.3.)

3.3.1 Cavitation Erosion

Expressed simply, cavitation is the local vaporization of liquid in low pressure regions of any liquid handling device. Cavitation, and the evolution of dissolved gases that often accompanies it, can affect pump performance (see NPSHR (net positive suction head required), Chapter 4) but does not damage the device. What can damage the device is the subsequent collapse of the vapor bubbles (but not the noncondensible gases) as they move into regions of higher pressure. The mechanism of damage is thought to be basically fatigue in nature, a consequence of high pressures and temperatures associated with bubble collapse.

Corrosive action has been suggested as a contributing factor but data are lacking. In a pump handling corrosive liquids, however, it is quite conceivable that the erosive action of bubble collapse could aid corrosion, as in erosion-corrosion, and there could be sufficient temperature rise in the region to accelerate the corrosion rate.

3.3.2. Fretting Corrosion

Minute, high frequency movement between contacting surfaces can produce deterioration of one or both of the surfaces. The phenomenon is known as fretting corrosion. It occurs in both wet and dry environments, and its occurrence in inert gases, noted by Faires [3.5], suggests it is not dependent upon electrochemical action.

3.4. Nonmetallic Corrosion

Nonmetallic materials used for chemical pump construction include rubber, plastic, ceramic, graphite, and glass. For practical purposes the last two materials are inert. Ceramic, too, is inert unless sintered with a metallic binder, in which case electrochemical action can affect the binder.

Rubber and plastic are both hydrocarbon polymers, thus subject to chemical action on their inter- and intramolecular bonds. Deterioration by the pumped liquid can occur in two modes:

Chemical attack—Action by the liquid alone causes deterioration of the polymer, e.g., softening or swelling.

Stress cracking—Tensile stress, either residual or imposed, *plus* the presence of a "stress-cracking agent," a liquid not normally corrosive to the polymer, causes cracking. As is the case with metals, even quite low concentrations of stress-cracking agents will promote failure.

Ultraviolet light also causes deterioration of polymers. While not a pumped liquid effect, UV degradation could be of consequence in pump installations exposed to strong sunlight.

References

3.1. L. H. Seabright and R. J. Fabian, "The Many Faces of Corrosion," *Mater. Des. Eng.*, pp. 85-91 (January 1963).

3.2. Z. D. Jastrzebski, *Nature and Properties of Engineering Materials*, Wiley, New York, 1959.

3.3. L. H. Van Vlack, *Materials Science for Engineers*, Addison-Wesley, Reading, Massachusetts, 1970.

3.4. A. Kratzer, "Corrosion and Erosion Damage to Chemical Centrifugal Pumps," in *Proceedings "Pump for Progress,"* 4th Technical Conference BPMA, April 1975. Bedford, UK.

3.5. V. M. Faires, *Design of Machine Elements*, 4th ed., Collier-Macmillan, New York, 1965.

4. System Characteristics

4.1. Pumping

Pumping involves the movement of liquid, or, occasionally, a liquid–gas mixture, from a suction source to a discharge point. Figure 4.1 shows a typical system and the hydraulic gradient associated with a particular flow through it.

4.2. Pump Energy

The first point to note from the hydraulic gradient is that the pump is the only device adding energy. And it has to add all the energy required; not only that to overcome the difference between pressures at the suction source and discharge point, but also the losses in the connecting conduits. While this point may seem trite, it is fundamental and cannot be overlooked. The energy added by the pump is equal to the system head or resistance.

4.3. Net Energy at Suction

Of equal importance to the pumping energy is the net energy available at the pump suction. Net energy means that over the liquid's vapor pressure at that point, and is shown in Fig. 4.1. To get the liquid into the pump without undue deterioration of performance or expected life, a pump requires a certain net energy at its suction. This net energy is commonly known as NPSH; it is dealt with in detail in Sections 4.8 and 4.9.

4.4. Flow

Pump size is determined by the required flow rate. For new plants or well-documented existing plants, flow rates are obtained from process data. When such data are not available, as might be the case in replacing an old pump, some means of "in field" flow measurement has to be used.

When the flow can vary, either increased to accommodate an upset or reduced for plant start-up or turndown, the various values should be specified. Conventional terms are:

Rated—Flow for which the pump is to be sized; usually the maximum flow.
Normal—Flow at which the pump will operate most of the time.
Minimum—Flow down to which the pump may operate; likely time at this condition to be specified.

Rated flow values often include some "margin" to cover uncertainties in process calculations or pump wear or both. To avoid oversizing, keep the margin small; 5% should be adequate if flow fluctuations have been correctly accounted for.

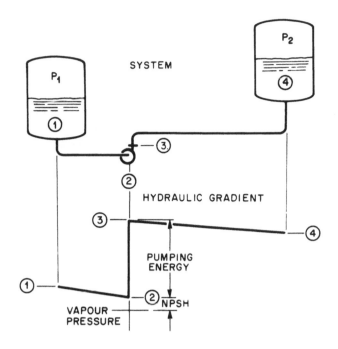

ENERGY LEVELS

① – EXIT FROM SUCTION SOURCE

② – PUMP SUCTION

③ – PUMP DISCHARGE

④ – DISCHARGE POINT

FIG. 4.1. Hydraulic gradient in a typical system. The pump must add all the energy, including losses in the conduits, to move the liquid from suction source to discharge point. The energy available at the pump suction over and above the liquid's vapor pressure is the NPSH available.

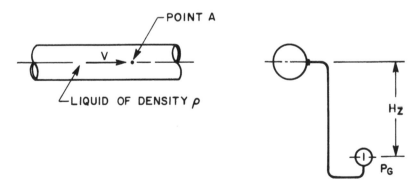

FIG. 4.2. Liquid flowing in a conduit. Total head at Point A is the static pressure plus the velocity head. The gauge indicates the static pressure in the conduit plus the pressure produced by elevation of the conduit above the gauge.

4.5. Energy Added

To produce the desired flow through a particular system, energy has to be added to the liquid (see the hydraulic gradient in Fig. 4.1). The energy necessary can be expressed in pressure or head units. A convenient way to illustrate the total energy of liquid and the interchangeability of pressure and head is to consider the conditions at a single point in a flowing conduit. (Figure 4.2 shows such a conduit.)

At Point A the static pressure, P_s, is that indicated by the gauge, P_g, less the correction for gauge elevation:

$$P_s = P_g - \rho g H_z \qquad (4.1)$$

The correction for elevation, $\rho g H_z$, takes account of the additional potential pressure applied to the gauge by the liquid leg between it and Point A. If the gauge were above the measuring point, the correction would be positive.

At Point A the liquid has a velocity, V, thus its total pressure, P_t, is the static pressure plus that produced by the velocity:

$$P_{\text{total}} = P_g - \rho g H_z + \frac{V^2}{2_g} \qquad (4.2)$$

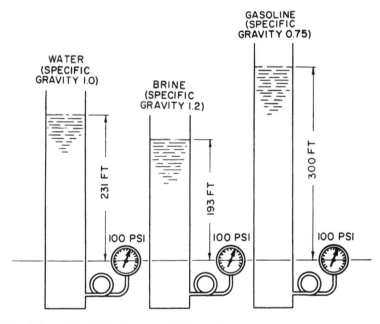

FIG. 4.3. Effect of liquid density on static head. Comparison of the heights of columns of water, brine, and gasoline needed to effect a 100 lb/in.²gauge pressure at the datum line.

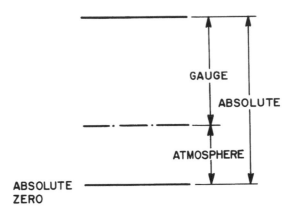

FIG. 4.4. Pressure terms. Gauge is the pressure above local atmospheric pressure and there-fore depends upon location and elevation. Absolute is referenced to absolute zero and is independent of location or elevation.

Equation (4.1) incorporates the general equation relating pressure to head:

$$P = \rho g H \tag{4.3}$$

Conversion from pressure to head and vice versa is made more tractable using specific gravity (SG) or relative density (RD). In United States units:

$$P = \frac{H}{2.31}(SG) \tag{4.4}$$

and in metric units ($P = $ kPa, $H = $ meters):

$$P = 9.81(H)(RD) \tag{4.5}$$

Figure 4.3 illustrates the relationship between pressure and the height or head of a liquid column for various SGs. Pressure at a point can be expressed in gauge or absolute terms. Gauge means over and above prevailing atmospheric pressure and therefore varies with location. Absolute refers to absolute zero and is independent of location. Figure 4.4 illustrates the concept.

4.6. System Characteristics

Careful establishment of system characteristics is crucial. Failure to do so can lead to the wrong pump being selected, resulting in problems with the process or equipment or both.

In most respects, system characteristics are essentially independent of the pump type. The one exception is NPSH where pulsating or fluctuating flows can have a major effect.

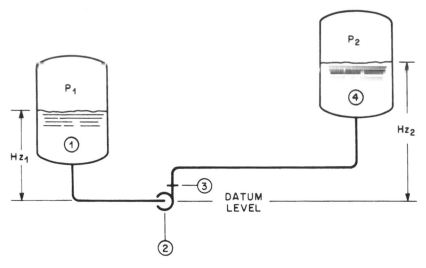

FIG. 4.5. Typical pumping system. Liquid is to be moved from a suction vessel at some elevation and pressure to a discharge vessel at another elevation and pressure.

4.7. System Head

Figure 4.1 shows the system head for a particular flow; the question now is how to determine it.

A general pumping system, less valves for simplicity, is shown in Fig. 4.5. The task is to pump from Vessel 1 to Vessel 2.

System head or resistance has three components: static pressure, elevation head, and friction loss.

Static pressure is the difference in vessel pressures; for Fig. 4.5 it is:

$$P_s = P_2 - P_1 \tag{4.6}$$

Elevation head is the difference in liquid level between the suction and discharge vessels. To avoid confusion, elevation head should be determined using a datum level. For horizontal pumps the datum level is usually the shaft centerline; for vertical pumps it is the centerline of the first stage impeller, except for can pumps where convention uses the centerline of the suction nozzle. A liquid level above the datum level is positive, one below is negative. For Fig. 4.5 the elevation head is

$$H_z = H_{z_2} - H_{z_1} \tag{4.7}$$

Friction loss in a system depends upon flow and the Reynolds number. The effect the Reynolds number has is on the variation of friction loss with flow. At values below the "transition" (consult the Hydraulic Institute tables [4.1]), flow is laminar and friction loss is proportional to flow; at values above the "transition," flow is turbulent and friction varies as the square of the flow ratio. The Reynolds number is a function of pipe size, liquid velocity, and liquid viscosity. For applications pumping liquids whose viscosity is close to water, the flow is typically turbulent. With higher viscosity liquids, the flow can be laminar and this should be checked by calculating the Reynolds number.

System friction covers losses from piping entrance and exit, fittings, valves, reducers, strainers, flow meters, and the piping run itself. For Fig. 4.5, all the losses are from (1) to (2) and from (3) to (4). Individual component losses are determined from the Hydraulic Institute friction tables or similar publications. Flow control valves, if used, require a certain minimum pressure drop to have command over the system. The value varies with valve type and has to be obtained from the valve manufacturer.

Putting the three components together yields the total system head or resistance. In head terms for centrifugal pumps:

$$H_{\text{Total}} = (P_2 - P_1)\frac{2.31}{SG} + (H_{z_2} - H_{z_1}) + H_{L_{1\text{-}2}} + H_{L_{3\text{-}4}} \tag{4.8}$$

Or in pressure terms, the practice for positive displacement pumps:

$$P_{\text{Total}} = (P_2 - P_1) + (H_{z_2} - H_{z_1})\frac{SG}{2.31} + (H_{L_{1\text{-}2}} + H_{L_{3\text{-}4}})\frac{SG}{2.31} \tag{4.9}$$

Figure 4.6 shows the various system head components and the resultant characteristic.

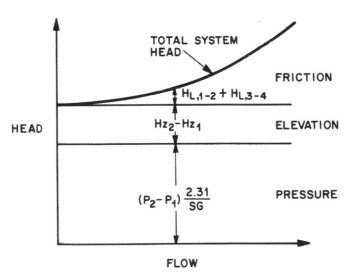

FIG. 4.6. System head. The total of pressure difference, elevation difference, and friction losses. Note that when the liquid is heated downstream of the pump, e.g., by an economizer, the pressure in the discharge vessel is converted to head using the SG of the pumped liquid.

Static pressure and elevation head are often independent of flow and can then be combined into a single term, static head. Figure 4.6 is drawn with this assumption.

In many instances the components of system head can vary with process conditions or time. Examples are static pressure varying with flow rate, elevation head changing with vessel levels (batch processing or system upsets), and friction loss characteristics being affected by liquid viscosity or pipeline condition (new to scaled). The extremes associated with these variations need to be determined to ensure pumping can be sustained under such conditions. Figure 4.7 shows such a system characteristic.

4.8. Hydraulic Power

Once the flow and corresponding system resistance have been established, the pumping hydraulic power can be calculated from

$$\text{hydraulic hp} = \frac{Q(H)(\text{SG})}{3960} \qquad (4.10)$$

where Q = flow (gal/min) and H = total head (ft), or

$$\text{hydraulic hp} = \frac{Q(\Delta P)}{1714} \qquad (4.11)$$

where ΔP = differential pressure (lb/in.^{2}).

The utility of hydraulic power is twofold. First, it enables an order of magnitude appreciation of the total pumping power. Second, it permits ready calculation of pump power once pump efficiency is known.

$$\text{pump} = \frac{\text{hydraulic hp}}{\eta} \qquad (4.12)$$

where η = pump efficiency (decimal), the ratio of power out over power absorbed.

4.9. NPSH

The fundamental definition of NPSH was given in the discussion on the basic pumping process. Pumps are liquid-handling machines. If the liquid, during its passage into the pump's working element, has its energy level lowered to the point where vaporization occurs, there are two possible consequences. First, there is the risk of cavitation erosion as the vapor bubbles move into

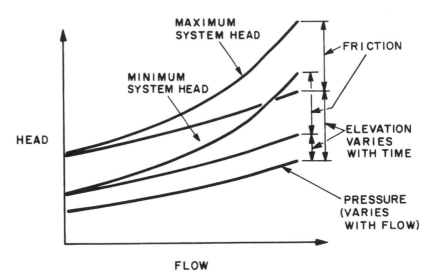

FIG. 4.7. Effect of variable static pressure and elevation head. As liquid is pumped into a gas filled, closed vessel or into the bottom of a tall reservoir, the static pressure or elevation head increases with time, giving a range of system heads over which the pump has to operate.

regions of higher energy and collapse or "implode." Second, the extent of vaporization may get to the point where there is a discernible reduction in pump performance. To avoid or minimize these ill effects, pumps require a certain amount of NPSH, giving rise to the term "NPSH Required" or NPSHR.

In centrifugal pumps, NPSHR is a product of kinetic action, thus it is independent of liquid density (or SG) and is always expressed in head terms. Reciprocating pumps, however, have valves whose opening is a dynamic action, making the dominant component of NPSHR a pressure. This consideration has led to the limited use of the term NPIP (net positive inlet pressure). Rotary pump NPSH requirements are essentially the product of kinetic action, but because they are positive displacement devices, they've followed the reciprocating pump convention of expressing NPSH requirements in pressure terms.

4.10. NPSHA

From pumping fundamentals (Fig. 4.1) and the above discussion, it is evident the NPSH required by the pump must be provided by the system. The term for this system property is "NPSH Available" or NPSHA.

For centrifugal and most rotary pumps, in which the flow is essentially constant, NPSHA is the difference between the total suction head and the liquid's vapor pressure, both expressed in head units, see Fig. 4.8. NPSHA calculations are usually made in absolute terms, since doing so avoids having to take account of atmosphere in the vapor pressure.

Using conditions measured at the pump suction, the rigorous equation for NPSHA is

$$\text{NPSHA} = \frac{(P_s + P_a)}{\rho g} + \frac{V^2}{2g} - \frac{P_v}{\rho g} \qquad (4.13)$$

The first two terms represent the total suction head (static head plus velocity head), and the third the head equivalent of vapor pressure.

At the plant design stage, conditions at the pump suction are determined from those in the suction vessel and a knowledge of the connecting conduit. Referring to the general system in Fig. 4.5, the NPSHA is

$$\text{NPSHA} = \frac{(P_{s_1} + P_a)}{\rho g} + H_{z_1} - H_{L_{1\text{-}2}} - \frac{P_v}{\rho g} \qquad (4.14)$$

or using Eq. 4.4 rearranged:

$$\text{NPSHA} = (P_{s_1} + P_a)\frac{2.31}{\text{SG}} + H_{z_1} - H_{L_{1\text{-}2}} - (P_v)\frac{2.31}{\text{SG}} \qquad (4.15)$$

For general guidance, the first term is the head equivalent of the pressure acting on the liquid surface, the second is the elevation of the surface from the datum level (figured +ve when above, −ve when below), the third is the friction loss from suction vessel to pump, and the fourth is the head equivalent of the vapor pressure. When the suction vessel is saturated, i.e., the liquid is at its vapor pressure, the first and fourth terms cancel, leaving

$$\text{NPSHA} = H_{z_1} - H_{L_{1\text{-}2}} \qquad (4.15)$$

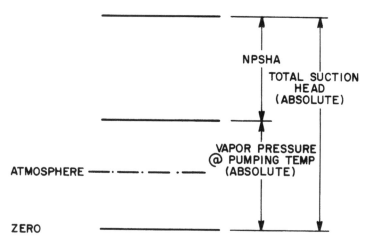

FIG. 4.8.   NPSHA. The energy available at the pump suction over and above the vapor pressure of the liquid, all expressed in head terms.

In making NPSHA calculations there is a tendency, born of conservatism, to build in "safety margins" by using the lowest conceivable vessel level or pressure. There is usually justification for some margin, but when it becomes excessive, as it can with "built in" or "hidden" margins, it can cause more harm than good. At best, the result can be larger and more expensive pumps than are necessary. With centrifugal pumps more can happen; the pump may suffer continual operating problems. See "NPSHR" in Centrifugal Pumps, Chapter 6, for a detailed discussion.

Equation (4.13) gives the NPSHA for steady-state conditions. When the flow can change rapidly or pulsates continually, the acceleration energy associated with such changes must also be provided by the system. The term used for this energy is "acceleration head" or H_a. Taking this into account, NPSHA becomes, for the system in Fig. 4.5,

$$\text{NPSHA} = (P_{s_1} + P_a)\frac{2.31}{\text{SG}} + H_{z_1} - H_{L_{1\text{-}2}} - H_a - (P_v)\frac{2.31}{\text{SG}} \qquad (4.16)$$

The magnitude of H_a depends upon the rate of flow change. That, in turn, depends upon the type of pump and the means of flow control, therefore determining H_a is one of the factors involved in refining the "hydraulic duty" once the pump type has been selected; see Fig. 1.1.

Reciprocating pumps, and some rotary pumps, produce a pulsating flow. The degree of pulsation depends upon the pump type and drive arrangement. The effects of pulsating flow on NPSHA are twofold. First, the acceleration head, H_a, discussed above, must be accounted for. Second, friction loss calculations should be based on the peak flow rather than the nominal.

Information for estimating H_a and allowing for peak flow friction is given in the Chapters 6, 7, and 8 for centrifugal, rotary, and reciprocating pumps, respectively.

4.11. Dissolved Gases

Dissolved gases coming out of solution as liquid enters a pump can cause a performance drop in the same manner as cavitation, i.e., by partial obstruction of the impeller's waterways. The phenomenon, however, is not the same as cavitation; the free gas is noncondensible, thus it will not "implode" as the ambient pressure increases. Without implosion there will not be any cavitation erosion. If anything, the presence of free gas will tend to reduce the likelihood of cavitation erosion (the gas acts as a shock absorber). Davidson notes the distinction between dissolved gas and cavitation. Both Penney and Tsai set down methods in Ref. 1.2 to allow for dissolved gas content in calculating NPSHA.

Reference

4 1. *Hydraulic Institute Pipe Friction Manual*, Hydraulic Institute, New York, 1961.

5. Pump Class Selection

The pump is one of the oldest devices known to mankind and is second only to the squirrel cage induction motor in numbers used. With a long history and wide usage, the pump has been subjected to sustained inventive ingenuity, resulting in its now being available in myriad types. To bring a sense of order to the profusion of types, the Hydraulic Institute [5.1] has published a classification chart of pump types; Fig. 5.1.

Even with a classification chart as an aid, selection of the most appropriate pump type for a particular duty can still be a daunting task. Just how daunting can be illustrated by observing that for many duties the choice of pump types is quite controversial, meaning there is likely more than one appropriate selection. This, of course, is how engineering should be, lest judgment be replaced by binary action.

The following material is intended to provide a broad guide to the selection of a pump class: whether centrifugal, rotary, or reciprocating. Once the class is chosen, the selection is refined by referring to the chapter dealing with that class.

Along with providing broad guidelines, this chapter gives some insight to the rationale behind the guidelines. Equipped with this, the reader is better able to judge the worth of the guidelines, and thus their relevance to a particular case.

Any selection process needs a sequence of decisions if it is to remain orderly. The sequence adopted for this volume is shown in Fig. 5.2.

The sole reason for employing a pump is to add energy to a liquid stream. Given this, the first choice should be based on hydraulic duty. Other conditions may dictate modifications to the hydraulic selection, but by modifying a sound hydraulic selection the chances of a poor final selection are reduced.

The hydraulic duty determined from process data in chapter 4 is the total for the system. For pump selection the hydraulic duty has to be that per pump, the pump rating. In the simplest case, one pump is used for the whole duty, the so-called "full capacity" pump. Sharing the flow between two or more pumps operating in parallel may be warranted when:

The flow is too large
The NPSHA is low
Operation includes sustained, wide flow swings
The required driver is too large

CLASSES OF PUMPS

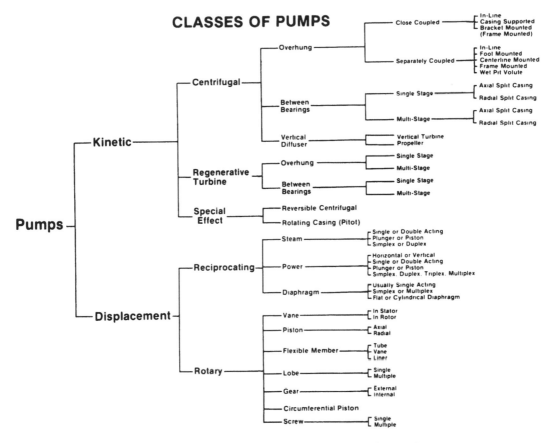

FIG. 5.1. Classes of pumps. (Reprinted with permission from Ref. 5.1.)

Similarly, sharing the energy increase between two or more pumps in series may be warranted when:

The energy increase is too high for a single pump
The NPSHA is low
The system head varies widely
The initial pressure is too high
The required driver is too large

Selecting the number of pumps often involves some iteration.

From Fig. 5.2 the pump rating first determines whether the pump is to be "kinetic" or "displacement." The distinction is in the pumping action and it has a major bearing on hydraulic performance. In "kinetic" pumps the liquid acquires energy by being accelerated to a high velocity, then has most of the velocity energy converted to pressure as the velocity is reduced to a usable value. "Displacement" pumps have quite a different action; they just "capture" a volume of liquid and move it into the process, all at usable velocities.

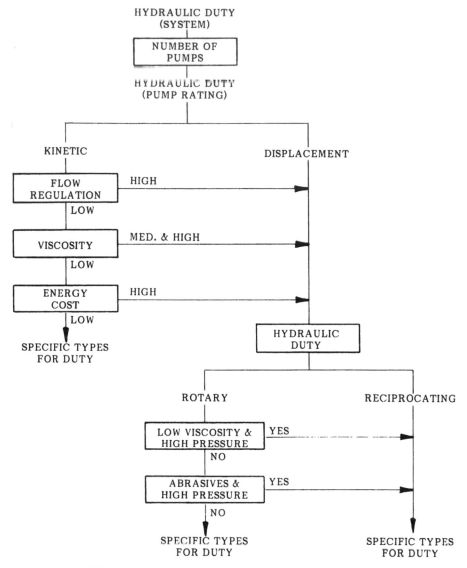

FIG. 5.2. Pump class selection on the basis of service conditions.

Being dependent upon attainable liquid velocities, kinetic pumps have a lower capacity limit (the passages become too small) and an upper energy limit (component erosion limits the velocity). Displacement pumps have an upper capacity limit (machine size dictated by inertia effects) but the energy that can be added is limited only by the mechanical strength of the particular configuration, or the drive capacity.

The choice for the hydraulic duty is easiest made by using a chart showing the appropriate upper limits of energy added (pressure rise) and flow for the various classes of pumps. Such a chart, presented by Krutzsch in Ref. 5.1, is shown in Fig. 5.3.

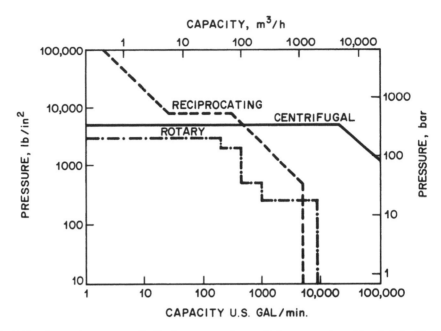

FIG. 5.3. Approximate upper limit of pressure and capacity by pump class. (Reprinted with permission from *Pump Handbook*, McGraw-Hill, New York, 1976.)

Where a chart such as Fig. 5.3 allows more than one class of pump, the choice is centrifugal, rotary, then reciprocating, depending upon factors other than the rated hydraulic duty. The first of these factors is flow regulation.

By the nature of their pumping action, kinetic and displacement pumps have distinctly different flow regulation. The energy added by kinetic pumps varies with through flow, hence their flow regulation is low (flow varies widely with system resistance). In displacement pumps the energy added depends upon system resistance while the mean flow remains practically constant. Their flow regulation is thus very high. Figure 5.4 illustrates the difference. If the service requires high flow regulation, a displacement pump is the appropriate choice.

The second factor is reassessing a kinetic pump selection is liquid viscosity. Kinetic pump performance deteriorates rapidly with increasing viscosity (see Chapter 6), and a displacement pump is usually the better choice whenever the liquid viscosity exceeds 500 SSU.

The final factor in kinetic versus displacement pumps has to do with energy consumption and its cost. For many applications, particularly those near the kinetic pump upper limit, displacement pumps are more efficient than the equivalent kinetic pump, thus consume less energy. With low cost energy the saving is not enough to offset the higher capital and usually higher maintenance costs of the displacement pump. When the energy cost is high, however, the balance can swing in favor of the displacement pump, and is therefore worth investigating.

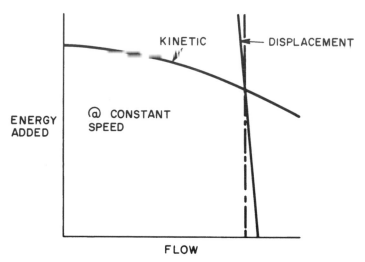

FIG. 5.4.   Flow regulation of kinetic versus displacement pumps.

Within the displacement pump group, the choice for hydraulic duty is confirmed by charts such as Fig. 5.3. Where both rotary and reciprocating pumps are allowed, the choice is rotary then reciprocating, subject to two general limitations.

Rotary pumps inherently have close running clearances, see Chapter 7. As liquid viscosity decreases, pump performance falls off and, if the liquid has low lubricity, clearance life can also deteriorate. When the pumped liquid has low viscosity (or low lubricity) and the pump differential pressure is high, a reciprocating pump is the more appropriate selection. As a guide, pumps for liquids whose viscosity is below 100 SSU warrant investigation.

All rotary pump designs, bar three, have a low tolerance of abrasive solids in the pumped liquid. The low tolerance is a product of their close internal clearances and the usual materials of construction; see Chapter 7 for further discussion. Given this limitation, reciprocating pumps are the preferred selection when liquids containing abrasive solids have to be pumped to pressures higher than 250–300 lb/in.^{2}gauge.

Economics, of course, also have a bearing on pump selection. In some instances it may be more economical to employ a "sacrificial" pump and replace it periodically rather than one whose service life will be significantly longer. Be sure, however, that all aspects of the "cost" of a "sacrificial" pump are taken into account, not just the lower purchase price.

With the class of pump chosen, the next step is to select a specific type within that class. Chapters 6, 7, and 8 deal with centrifugal, rotary, and reciprocating pumps, respectively. Chapter 9 covers metering, a duty readily identified within the various process requirements. Chapter 10 deals with the various pump types used in utility services.

Difficulty of sealing and liquid toxicity are two factors of major importance in pump selection. Their significance, however, comes to bear in the choice of type within a class rather than the class itself.

Reference

5.1. *The Standards of the Hydraulic Institute*, 14th ed., Hydraulic Institute, Cleveland, Ohio.

6. Centrifugal Pumps

6.1. Basic Arrangement

Figure 6.1 shows a rudimentary end suction centrifugal pump; its basic parts and their functions are:

impeller	imparts energy to the liquid by the action of its vanes; is the *only* pump component that adds energy to the liquid
inducer (optional)	low head, low NPSHR booster impeller; provides the NPSH required by the impeller
clearance	restricts leakage of high energy liquid back to the impeller inlet
casing	guides liquid to the impeller; collects liquid from the impeller and slows (diffuses) it to a usable velocity; primary element of pressure boundary; incorporates nozzles; may incorporate pump feet
cover	closes casing (second element of pressure boundary); supports the bearing housing
seal	seals the opening where the shaft passes through the pressure boundary
shaft	drives and supports the impeller
bearings	supports the rotor (impeller plus shaft)
housing	supports the bearings; retains bearing lubricant; keeps lubricant clean

6.2. Performance Characteristics

A typical centrifugal pump performance characteristic is shown in Fig. 6.2. Total head (energy added), power absorbed (for a particular SG), and NPSHR (net energy required at the inlet) are plotted against flow. These are the

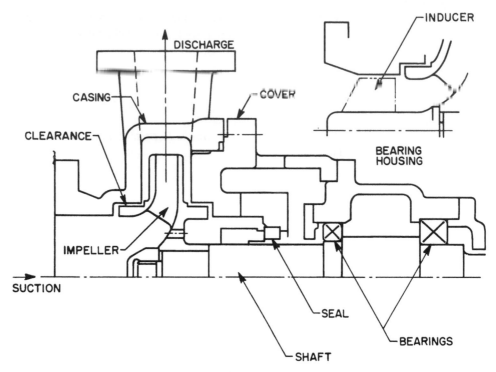

FIG. 6.1. Centrifugal pump; basic components.

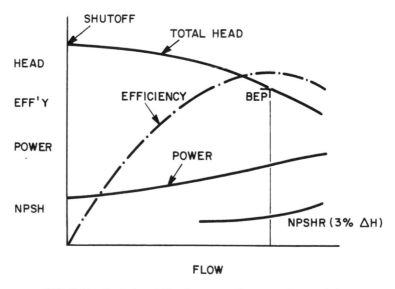

FIG. 6.2. Typical centrifugal pump performance characteristic.

working characteristics of the pump. Pump efficiency (derived from flow, total head, and power, hence the phantom line) is also plotted against flow to locate the best efficiency point (BEP) and indicate the pump's most effective operating range.

6.3. Specific Speed

Centrifugal pumps are produced in a very wide range of hydraulic designs. To categorize these designs, two concepts are used. The first of these is specific speed, designated N_s.

Derived from similarity considerations, specific speed is a number that broadly defines a pump's impeller geometry and performance characteristic regardless of size. The equation is

$$N_s = \frac{\text{RPM(total flow)}^{0.5}}{(\text{head/stage})^{0.75}} \tag{6.1}$$

In its original form N_s was dimensionless, but conventional usage with convenient units requires that the units be identified (either US gal/min and ft or m³/h and m). N_s is calculated for performance at BEP with maximum diameter impeller. The general variation of impeller geometry with specific speed is shown in Fig. 6.3. The geometry of an impeller has a bearing on its head and power characteristics, and consequently on its efficiency. Figure 6.4 shows how performance characteristics vary. Figure 6.5, from Fraser and Sabini [6.1], gives values of peak efficiency for pumps of various specific speed and capacity.

Appreciating how a centrifugal pump's head and power characteristics vary with specific speed is important. From Fig. 6.4 the following can be noted:

The head rise to shutoff increases with specific speed. At low specific speeds the head characteristic is flat or even drooping (maximum head at some flow greater than at zero), while at high specific speeds the characteristic is steep, with shutoff head 2–3 times that at BEP.

The power characteristic changes from positive (power increasing with flow) to negative as specific speed increases. Because the power characteristic changes inclination, there is a small range of specific speeds where the power characteristic peaks in the region of BEP. Such a characteristic is known as "non-overloading."

The typical head and power characteristics are consistent with attainable efficiency. Other characteristics are possible but generally at some expense to efficiency. As an example, constantly rising head and non-overloading, two "safe" characteristics, can be provided outside their usual ranges. To do so,

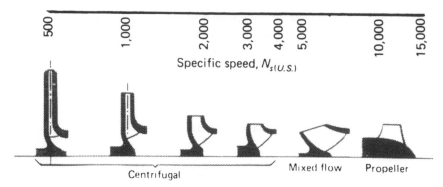

FIG. 6.3.　Impeller shape versus specific speed.

however, the impeller will have to be larger than "normal," thus increasing the power lost to disc friction and lowering pump efficiency.

Calculating the specific speed for a particular duty, assuming operation at BEP, gives a clue to the feasibility of a centrifugal pump for the duty and enables an estimate of its power.

6.4. NPSHR

Chapter 4 noted a pump's need for a certain net energy at its suction. The term for this need is NPSH required or NPSHR. As outlined in Chapter 4, if this need is not satisfied by the system, the pump will suffer damage (cavitation erosion) or a loss of performance or both.

In current practice, a centrifugal pump's NPSHR characteristic is determined by a discernible drop in performance. The method used is to run the pump at constant capacity, reducing the NPSHA until the total head drops. NPSHR is the NPSHA at a particular head drop, 3% by most standards. A typical NPSH test plot is shown in Fig. 6.6. The NPSHR characteristic is developed by running this test at four or more capacities.

Basing NPSHR on a certain degree of performance drop is a sound, practical approach, but does have two limitations. First, the degree of performance drop is dependent upon the ratio of vapor to liquid volumes (v_g/v_f)

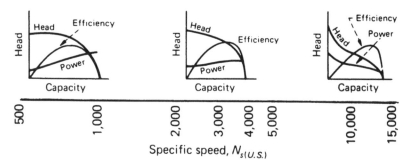

FIG. 6.4.　Typical performance characteristic variation with specific speed.

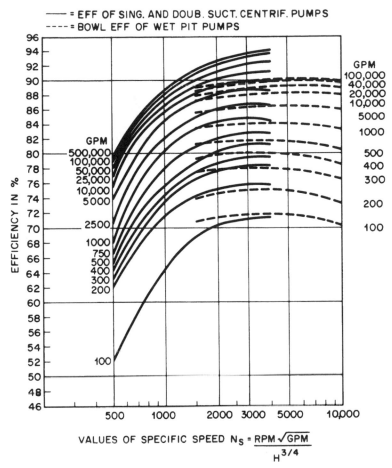

FIG. 6.5. Pump efficiency versus specific speed.

at the pump's suction conditions. For the same suction pressure, a liquid whose v_g/v_f is high will cause a greater performance drop than one whose v_g/v_f is low. Thus, for the same degree of performance drop, NPSHR decreases with v_g/v_f. NPSH test plots for two liquids, cold deaerated water, and a liquid of lower v_g/v_f, are shown in Fig. 6.7. The Hydraulic Institute publishes a chart for the so-called "hydrocarbon correction" of NPSHR; see Fig. 6.8. Most plant designers ignore the correction, preferring to leave it as an extra safety margin.

The second limitation has to do with the extent of cavitation. At 3% head drop, cavitation within the impeller has, by definition, developed to the extent that it is causing a 3% reduction in total head. Depending upon the liquid pumped and the impeller's material, design, and energy level, cavitation to this extent, and even less, can be sufficient to cause cavitation erosion. The discovery of this fact led to the concept of lowering the degree of performance drop used to define NPSHR. Further investigation. e.g., that by Vlaming [6.2] and Schiavello [6.3], has shown that cavitation erosion can occur at NPSH values well above those for any discernible head drop, thus

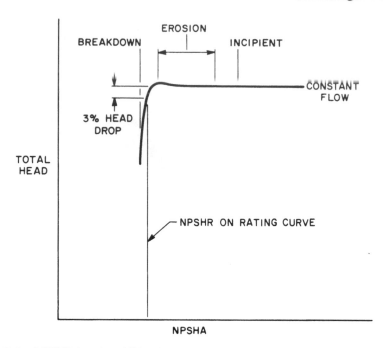

FIG. 6.6. NPSHR based on 3% head drop. Note possible erosion with NPSHA above value based on 3% head drop.

negating the concept of less allowable head drop. Figure 6.9 shows the various effects of decreasing NPSHA superimposed over head at constant flow. The practical solution to the dilemma created by erosion with ostensibly adequate NPSHA is the adoption of plant design and pump selection practices based on those factors found to have a bearing on damage. A major factor in determining the necessary practices is the concept of suction specific speed, the second concept used to categorize centrifugal pump hydraulic designs.

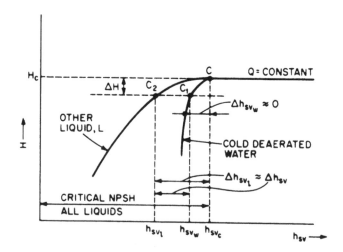

FIG. 6.7. Cavitation tests with different liquids at constant speed and constant capacity. (Reprinted with permission from Ref. 1.1.)

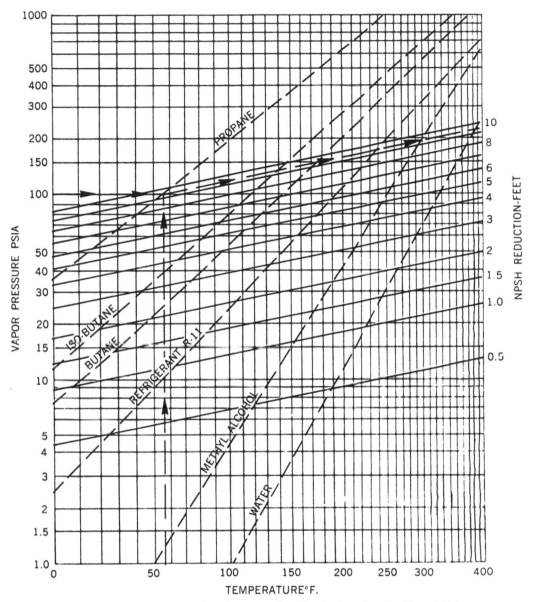

FIG. 6.8. NPSH reductions for pumps handling hydrocarbon liquids and high temperature water. (Reprinted with permission from Ref. 5.1.)

6.5. Suction Specific Speed

Drawing from specific speed, Wislicenus et al. [6.4] in 1939 advanced the concept of suction specific speed, S, to define an impeller's suction geometry and performance regardless of size. The equation is

$$S = \frac{\text{RPM(total flow)}^{0.5}(\text{flow/eye})^{0.5}}{(\text{NPSHR}_3)^{0.75}} \tag{6.2}$$

As with N_s, conventional usage require that the units be identified. The subscript "3" for NPSHR indicates a 3% head drop, the criterion for which most correlations of S have been developed.

From the point of view of system design and pump selection, the significance of S is twofold: First, as S increases so does the separation between NPSHR for a 3% head drop and that for incipient cavitation. Wislicenus [6.5] noted this with the observation that significant cavitation was almost inevitable, along with a drop in efficiency, when S exceeded 10,000 (US gal/min). Second, the onset of secondary flow at the inlet (suction recirculation), as a fraction of best efficiency capacity, increases with S. Fraser [6.6] provided the first relationship connecting S with the inception of suction recirculation. Figure 6.10 shows the combined result of these two effects on NPSHR characteristics. The net result is that as S increases, there is a greater risk of noise, vibration, and impeller erosion from either classical cavitation or suction recirculation. The risk is compounded by broadening the pump's operating flow range.

Most industry specifications now limit S to 11,000 or 12,000 (US gal/min). While better than nothing, single number limits are often compromises. Lobanoff and Ross [6.7] present a chart giving "stable" flow ranges for various values of S; see Fig. 6.11. Figure 6.12 shows "good" values of S versus N_s for the three usual impeller configurations. Doolin [6.8] and Ross and Fabeck [6.9] make the point that an inducer of S between 15,000 and 17,000 has the same rangeability as an impeller of $S = 11,000$. Pump "size" is important too. While not yet well defined, there seems little likelihood of serious damage to pumps whose power at BEP is less than 100 hp.

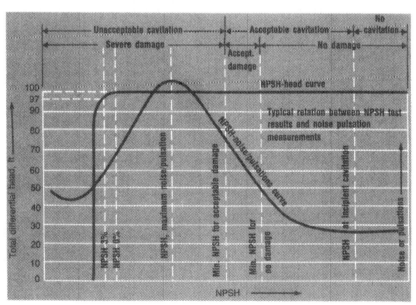

FIG. 6.9. Effect on erosion, noise, and pressure pulsations as NPSHA is decreased. (Reprinted with permission from *Oil & Gas Journal*, November 19, 1984.)

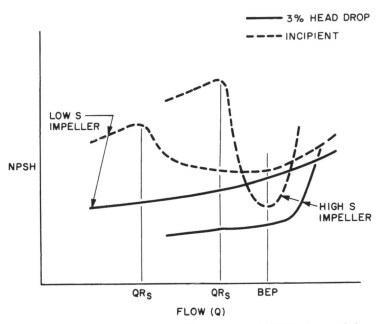

FIG. 6.10. Effect of suction specific speed, S, on NPSHR characteristic.

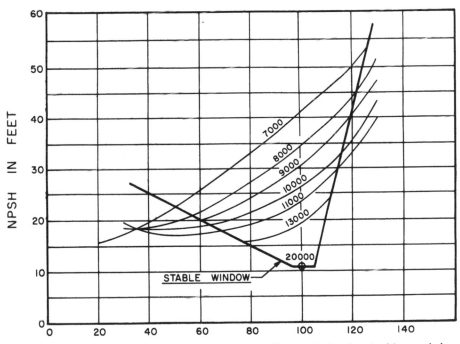

FIG. 6.11. "Stable" flow range versus suction specific speed. (Reprinted with permission from Ref. 6.7.)

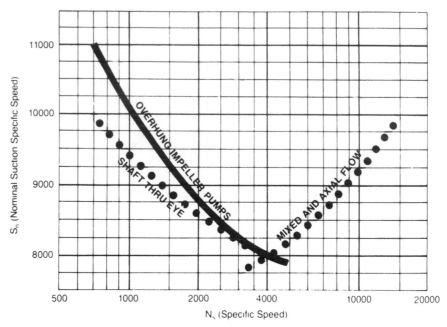

FIG. 6.12. Relationship between specific speed and suction specific speed. (Courtesy Worthington Pump, Dresser Industries, Inc.)

6.6 Number of Stages

The rudimentary pump shown in Fig. 6.1 develops all its head in a single stage. In Chapter 5, note was made of an upper limit to the head that can be developed in one stage; that limit was based on liquid velocity and component erosion. With appropriate materials, freedom from active corrosion, and clean liquids, this limit is high, of the order of 2500–3500 feet, higher in small pumps; see 6.20.1.8, High Speed Single Stage Pumps.

With the concepts of specific speed and suction specific speed now available, other limitations are evident. First, pump efficiency deteriorates rapidly at specific speeds below 1000 (US gal/min), and is considered the lowest tolerable at specific speed 500; see Fig. 6.5. Figure 6.3 provides some explanation: the impeller geometry becomes harder to manufacture as specific speed decreases. Clearly, then, specific speed should be kept above 500, and if minimum energy consumption is important, above 1000.

As the head required of a stage is increased, the specific speed decreases; see eq. (6.1). One means of compensating for this and maintaining a "good" specific speed is to increase the rotative speed. While doing this has the desired effect on specific speed, it also increases the suction specific speed, hence the NPSHR; see eq. (6.2). Since it is not always feasible to provide the additional NPSH, maintaining specific speed by increasing the rotative speed is limited by the effect on NPSHR.

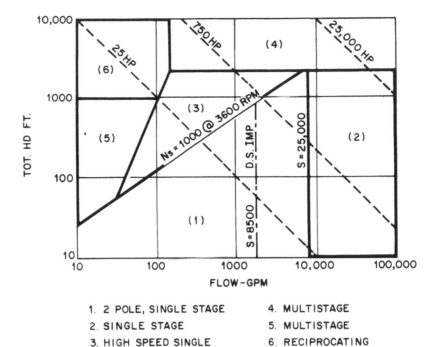

FIG. 6.13. Pump type selection based on energy conservation. (Courtesy Worthington Pump, Dresser Industries, Inc.)

The problem now becomes one of devising a pumping arrangement that offers good efficiency at acceptable NPSHR. Examination of eq. (6.1) shows this can be achieved by developing the required head over more than one stage, thus lowering the head per stage to maintain the desired specific speed. Generally a multistage pump (see 6.20) is used to develop the required head. Occasionally, when service conditions dictate, individual pumps in series are employed.

Doolin [6.10] addressed the problem of centrifugal pump selection with the objectives of minimum energy consumption and tolerable NPSHR, and as a result developed the selection chart given in Fig. 6.13.

6.7 Operating Capacity

The operating capacity of a centrifugal pump is determined by the intersection of its head capacity curve and the system head curve, see Fig. 6.14. For the pump performance to be consistent with its published curve, the NPSH available from the system has to exceed that required by the pump at the capacity given by the pump/system head intersection. Figure 6.14 illustrates the point. If the NPSHA is less than that required, and this, too, is shown in Fig. 6.14, the pump will still operate at the pump/system head intersection, but the pump head will now be modified by the effect of cavitation.

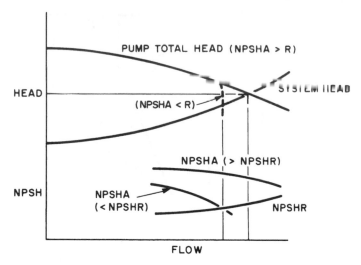

FIG. 6.14.   Pump and system interaction.

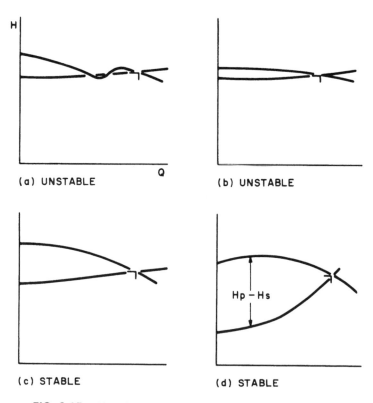

FIG. 6.15.   Unstable and stable pump and system combinations.

6.8 Stability

Stable operation requires that there is one, well-defined intersection between the pump and system head characteristic. The most common means of trying to insure this is to specify a constantly rising (with decreasing flow) head characteristic, the so-called "stable" characteristic. Doing this can preclude many entirely useful pumps and overlooks that stability is a product of pump/system interaction, not just a pump property.

Figure 6.15 shows four pump/system head combinations. The first two, (a) and (b), are classified "unstable." In (a) the pump characteristic has a "dip" which results in three possible interactions with the system characteristic. Combination (b) involves a pump whose characteristic meets the "stable" requirement noted above, but has such an acute intersection with the system characteristic that flow control will be very difficult.

The other two combinations in Fig. 6.15, (c) and (d), are classified "stable." The pump head characteristic in (c) is sufficiently steep, at least 10% rise to shutoff, to give a controllable intersection with the "flat," predominately static, system head. Combination (d) is a "drooping" or "unstable" pump characteristic against a "steep," predominantly friction, system characteristic. There is one, well-defined intersection, and provided the difference between pump and system head, $H_p - H_s$, is greatest at zero flow, the pump can be controlled at all flows.

6.9. Combined Characteristics

For the reasons given in Chapter 5, pumps may be arranged to operate in either parallel or series (and, on occasion, in series and parallel). The head characteristics for the two combinations of centrifugal pumps are determined as follows.

6.9.1. Parallel Operation

Pumps in parallel have a common suction and discharge, thus operate at the same total head. Total flow is therefore determined by adding flows at equal heads; see Fig. 6.16. Two further points of note are shown in Fig. 6.16. First, Pump B will not deliver any flow until Pump A is delivering some 35% of its best efficiency flow. The import of this is that pumps should not be operated in parallel in regions where the head characteristic of one or both is relatively flat. Second, Pump B operating alone is running at around 130% of best efficiency flow, a flow at which the NPSHR or power may be higher than expected.

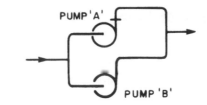

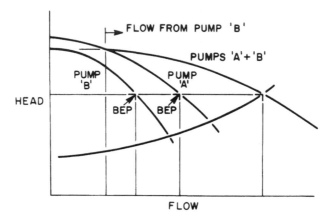

FIG. 6.16.   Parallel operation of centrifugal pumps.

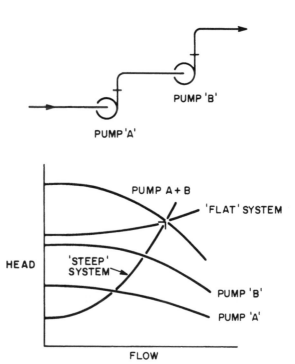

FIG. 6.17.   Series pumping.

6.9.2. Series Operation

Pumps in series handle the same flow unless there is some intermediate bleed-off. The combined head characteristic is therefore determined by adding heads at equal flows; see Fig. 6.17. When the system head is "flat," there will not be any flow until all the pumps are on line. A largely friction system, however, will likely have increasing flow as each pump is brought on-line. Figure 6.17 shows the effect of the two system characteristics. Pressure containment of both the casing and the seal needs care when considering series pumping, and should be checked for both rated and inadvertent zero flow operation.

6.10. Adjusting Pump Performance

A centrifugal pump is capable of a range of performance, not just the single head capacity characteristic shown in Fig. 6.2. Performance is adjusted (permanently) by reducing the impeller outside diameter in radial, Francis, and mixed flow impellers, see Fig. 6.3, and by changing the vane angle in axial flow impellers. The range of performance adjustment possible by impeller diameter alone decreases with increasing specific speed.

Performance variation with impeller diameter follows what are known as the "affinity laws." These say flow varies directly with the diameter ratio, head varies as the square of the diameter ratio, and power varies as the cube. Expressed as equations:

$$\frac{Q_2}{Q_1} = \frac{D_2}{D_1}; \quad \frac{H_2}{H_1} = \left(\frac{D_2}{D_1}\right)^2; \quad \frac{P_2}{P_1} = \left(\frac{D_2}{D_1}\right)^3 \tag{6.3}$$

where the subscripts 1 and 2 refer to the original and new conditions, respectively.

Figure 6.18 shows the head capacity characteristic of a particular pump with impeller diameter D_1. The pump's performance is to be adjusted to meet a duty Q_2, H_2, by reducing its impeller diameter.

Combining the Q and H terms of Eq. (6.3) yields the relationship

$$\frac{H_2}{H_1} = \left(\frac{Q_2}{Q_1}\right)^2 \tag{6.4}$$

or in words: Head varies as the square of the flow ratio. Picking one or two arbitrary flows, the point Q_2, H_2 can be extrapolated so it crosses the head characteristic for D_1. Connecting the points gives an intersection with the D_1 characteristic, the point Q_1, H_1. Using the ratio of heads since it's easier to read accurately, the "theoretical" impeller diameter ratio is given by

$$\frac{D_2}{D_1} = \left(\frac{H_2}{H_1}\right)^{\frac{1}{2}} \tag{6.5}$$

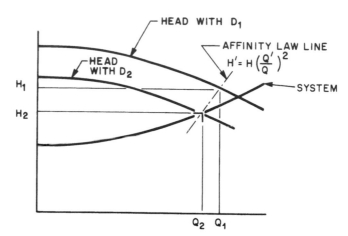

FIG. 6.18.   Adjusting pump performance by changing impeller diameter.

The diameter ratio in Eq. (6.5) is "theoretical" because the actual performance change will be slightly greater than that given by the "affinity laws," a consequence of the design being changed as the impeller diameter is reduced. To compensate, the "theoretical" diameter ratio is modified by an empirical factor to give the "actual." Figure 6.19 gives one such factor, which is applicable for impellers of N_s up to 2500 (US gal/min basis). For higher specific speeds, consult Ref. 1.1 or the manufacturer.

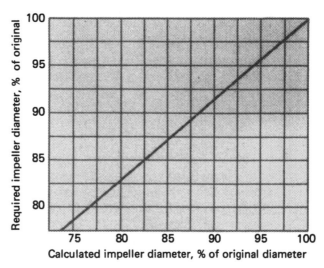

FIG. 6.19.   Correction factor for actual impeller cutdown (for specific speeds up to 2500, US gal/min basis.)

Reducing the impeller diameter can increase NPSHR, see Ref. 6.11, and raise the recirculation capacity, see Ref. 6.6. The likelihood of such problems increases with specific speed and is the reason why the allowable impeller diameter reduction decreases with increasing specific speed. For a given specific speed, the particular design also has a bearing on NPSHR and recirculation, those of high suction specific speed and efficiency tending to be more vulnerable.

6.11. Varying Capacity

In many processes it's desirable to have a pump whose rated capacity is greater than the normal requirement. One of the virtues of the centrifugal pump is that its capacity can be easily varied by changing either the system or pump head characteristic. Changing the system head is the more common approach and involves throttling to increase the friction head component, thus decrease the flow. Figure 6.20 illustrates the approach. While throttling usually reduces pump power, the additional friction introduced represents an additional energy loss to pump that capacity.

For small system flows, a modulating bypass may be the more convenient way to change the system head. With this arrangement, see Fig. 6.21, maximum flow to the process occurs with the bypass closed. As process demand falls, the bypass is opened, which has the effect of increasing the pump's gross flow while lowering that to the process. To appreciate why this is the case, note that opening the bypass effectively creates two systems in parallel, whose resultant characteristic is found by adding flows at equal heads. Operating against a lower total system characteristic, the pump's flow is higher and the

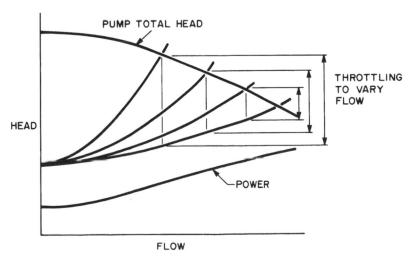

FIG. 6.20. Varying pump flow by throttling.

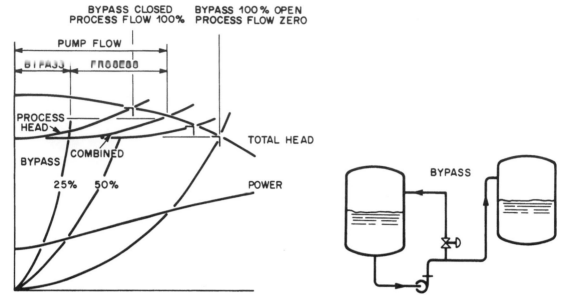

FIG. 6.21. Varying "net" pump flow by bypassing.

head lower. At the lower head the flow to the process is lower; see Fig. 6.21. Bypassing results in increasing pump power while the system requirement is decreasing, but it does avoid reducing pump flow to very low values with attendant pump and control problems.

Changing the pump characteristic can be done in three ways: changing the impeller diameter, the rotative speed, or the NPSHA. The first approach is more for adjusting rated performance and has already been discussed. The third approach is what is known as "submergence control," in which the liquid level in a saturated suction vessel controls the flow. It is a rarely used but viable technique for low NPSH values, say up to 6 ft.

Changing the rotative speed is the most common way of varying the pump characteristic. The variation of pump performance with speed also follows the "affinity laws:"

$$\frac{Q_2}{Q_1} = \frac{N_2}{N_1}; \quad \frac{H_2}{H_1} = \left(\frac{N_2}{N_1}\right)^2; \quad \frac{P_2}{P_1} = \left(\frac{N_2}{N_1}\right)^3 \tag{6.6}$$

The rotative speed for another point on a system head is found using the same basic technique as that for determining a new impeller diameter. For changes in speed, however, the affinity laws hold with high accuracy for speed ratios $N_2/N_1 \geq 0.5$, so there is no need for a "correction factor."

Variable speed pump operation is illustrated on Fig. 6.22. The approach has two advantages over throttling control: the operating point remains close to BEP, and the pump energy is only that necessary for the flow. Provided the means of varying the speed does not incur a high loss, the second advantage

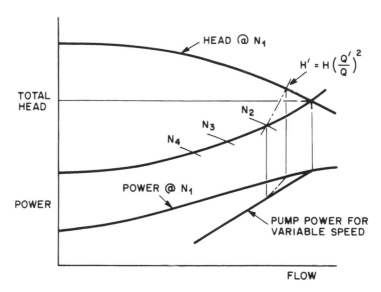

FIG. 6.22. Varying flow by changing speed.

can yield a substantial energy saving when operating over a wide flow range. In this connection, variable frequency drive (VFD) is a notable advance; see Ref. 6.13.

NPSHR is generally assumed to vary with speed in the same manner as head, i.e., as the square of the speed ratio. Stepunoff [6.12(a)] has presented data to the contrary, and Yedidiah [6.12(b)] has suggested the following:

$$\frac{\text{NPSHR}_2}{\text{NPSHR}_1} = \left(\frac{N_2}{N_1}\right)^{1.7} \tag{6.7}$$

6.12. Effects of Varying Flow

As a centrifugal pump is run over its range of possible flows, the design is no longer matched and various hydraulic events take place. These events occur in all designs. Whether they're of sufficient intensity to limit the flow range depends upon the energy density of the pump (nominally energy/rotor weight), its detail design and, to some extent, the liquid pumped.

Figure 6.23 shows the five hydraulic events most likely to affect a pump's allowable flow range. Some detail on each is given below:

1. Impeller Inlet Choking. The area between the vanes is no longer large enough for the flow, causing the NPSHR characteristic to steepen sharply. Inlet choking varies with suction specific speed (see discussion in this chapter), being typically 130% of BEP for low S impellers and falling to 105% as S increases; see Fig. 6.11.

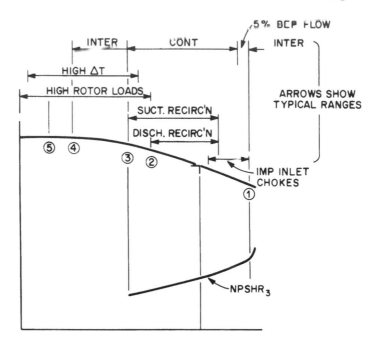

FIG. 6.23. Effects of varying flow on centrifugal pump behavior.

2. Discharge Recirculation. Below the discharge recirculation capacity, secondary flows exist at the impeller discharge. These can cause discharge vane erosion (in the vane/shroud fillet), impeller shroud fracture, and fluctuating axial thrust (often discernable as "rotor shuttle"). There is a possible connection between discharge recirculation and fluctuating radial thrust. The onset ranges from 70% of BEP (for conservative designs of lower efficiency) to beyond BEP in extreme designs.

3. Suction Recirculation. At the suction recirculation capacity a reverse flow is established in the outer region of the impeller eye. If the reverse flow is intense enough, it can cause flow surging, vibration, cavitation-like noise, and impeller vane erosion (from the upper side). The onset of suction recirculation is closely related to suction specific speed—see discussion in this chapter—and ranges from 50% of BEP in low S designs to beyond BEP for high S designs.

4. Rotor Loads. Depending upon both the hydraulic and mechanical design, rotor loads, either static or dynamic, can reach a point where the resultant rotor deflection is high enough to cause rapid clearance and seal wear, perhaps even premature shaft and bearing failure. Flow limits based on rotor loads range from none to around 50% of BEP. The source of rotor loads is a complex subject and is given further attention later in this section.

5. Temperature Rise. In a centrifugal pump most of the difference between input and output energy goes into heating the pumped liquid. As flow decreases, the resultant temperature rise can be sufficient to pose the risk of flashing within the pump. The temperature rise of concern is not so

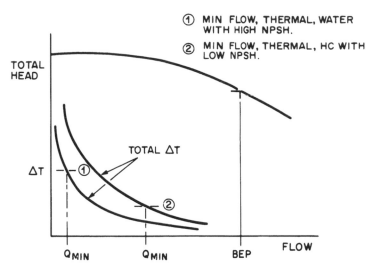

FIG. 6.24. Temperature rise characteristic and NPSH margin determine thermal minimum flow.

much that to a point of high pressure, since the pressure there will prevent flashing, but that in leakage back to suction pressure. Because such leakage has been pumped, then had its head broken down, its total temperature rise is approximated by

$$\text{Total } \Delta T = \frac{H}{778 \eta C_p} \tag{6.8}$$

where H = head in leakage flow, η = efficiency as a decimal, and C_p = specific heat. The "thermal" flow limit can range from only a few percent of BEP pumping water with high NPSH to near normal flow pumping a hydrocarbon with minimal NPSH margin; see Fig. 6.24.

6.13. Allowable Flow Range

In a general sense a pump has two operating regions: continuous and intermittent. Continuous is the region over which the pump can be run and still realize design life. Intermittent is that where operation is tolerable but if prolonged will significantly shorten service life. Figure 6.25 illustrates the idea by showing, in a qualitative manner, how pump wear rate varies with flow.

Continuous flow limits are usually maintained by pump selection and flow control. To insure the pump is not operated below the intermittent region, a bypass, sized for the lower limit, is necessary. The bypass can be fixed, manual, or automatic; see Chapter 15. When the difference between the minimum intermittent and continuous flows is small or extended operation

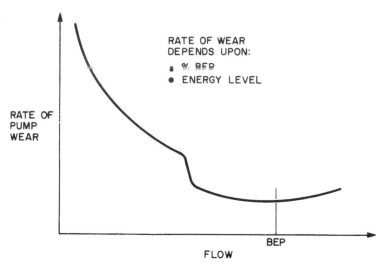

FIG. 6.25.   General effect of operating flow rate on pump wear.

below the minimum continuous flow is expected, the bypass is sized for the minimum continuous flow.

Determining working limits is not easy. First, the "off design" behavior of centrifugal pumps is still far from an exact science. Second, the onsets of the various events overlap, so what is limiting for one pump or service may not be for the next. Figure 6.23 shows typical ranges of onset for the five events to try to illustrate this. Some general comments are:

1. The maximum intermittent flow shouldn't exceed the capacity where the NPSHR curve steepens. A margin of say 5% of BEP flow then gives the maximum continuous capacity.
2. A good starting point for the minimum continuous flow is the suction recirculation capacity. For pumps handling hydrocarbon, lower the minimum continuous flow to 60% of suction recirculation capacity. For conventional one-and two-stage pumps smaller than 100 hp/stage, the minimum continuous and intermittent flows are the same.
3. Rotor loads or temperature rise usually determine the minimum intermittent flow. If the temperature rise limit exceeds the minimum continuous flow from two, it becomes the minimum continuous flow.
4. In high speed, single stage, and multistage pumps, both the minimum continuous and intermittent flows are often determined by rotor loads.
5. On occasion, head rise for stable operation, driver power (in high specific speed pumps), or the need to handle some entrained gas (see discussion later in this chapter) will determine the minimum bypass flow.

6.14. Radial Loads

In a centrifugal pump, radial loads of hydraulic origin can take three forms:

1. Steady radial thrust produced by the distorted pressure distribution around a single volute at off-design operation, or by the pressure difference across an assymetric twin volute, or around an eccentric diffuser. Agostinelli et al. [6.14] provided classical data on single volutes. Gasuinas [6.15] reported the ill effects of assymetric twin volutes. Hergt and Krieger [6.16] noted the effect of various eccentricities in diffusers.
2. Fluctuating radial load, probably produced by pressure pulsations associated with secondary flows at the impeller discharge (discharge recirculation). According to Uchida et al. [6.17], the magnitude of fluctuating radial loads, superimposed on the steady load, can be as high as 50% of the steady load.
3. Low frequency rotating force produced by diffuser stall. Hergt and Krieger [6.16] and Kanki et al. [6.18] provide some data on this phenomenon and the loading it produces.

6.15. Effect of Viscosity

In a kinetic machine such as a centrifugal pump, increasing the viscosity of the pumped liquid has two effects: it increases the friction losses, hence the power, and the thicker boundary layers alter the effective geometry, thus lowering the total head. With higher power and less head, the overall result is a substantial drop in efficiency, being greatest at low specific speeds. Figure 6.26 shows how increasing the viscosity from 32 SSU lowers the performance of a particular pump. Charts published by the Hydraulic Institute are used to estimate pump performance with liquids more viscous than water. Unless there are compelling reasons to do otherwise, centrifugal pumps should not be used for liquids whose viscosity exceeds 500 SSU.

6.16. Entrained Air or Gas

As already noted in Chapter 4, entrained air or gas can affect pump performance in the same manner as cavitation. Figure 6.27 shows the effect as gas is increased from 1 through 6% by volume; the total head is reduced and the minimum flow necessary to prevent gas binding is increased. Grohmann [6.19] reports inducer pumps successfully handling liquids with

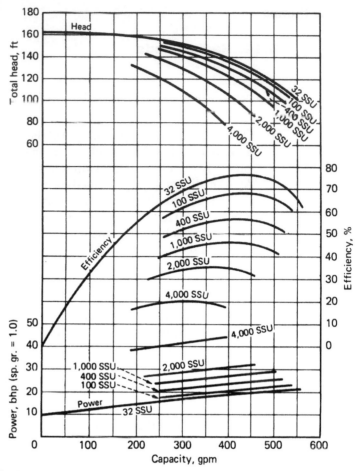

FIG. 6.26. Effect of pumped liquid viscosity on centrifugal pump performance.

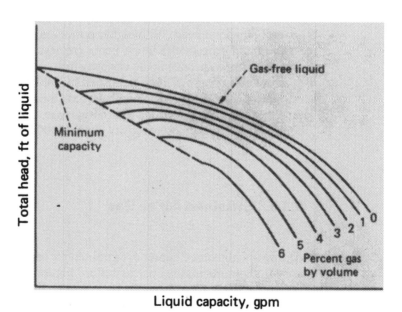

FIG. 6.27. Effect of entrained air or gas content on performance of conventional centrifugal pump.

gas content approximately 50% by volume. This capability is attributed to the inducer's lower vane pressure differential reducing the amount of gas separation, hence the blockage it causes.

6.17. System Corrections

6.17.1. Friction Losses

For the purposes of system calculations, centrifugal pumps produce pulsation-free flow, thus there is no need to account for peak flow in friction calculations.

6.17.2. Acceleration Head

The only circumstances where acceleration head is a factor in centrifugal pump NPSHA are when the pump has a "long" suction line or its flow is controlled by a quick acting valve.

With a "long" suction line there is a risk the pump can accelerate so quickly on starting that the suction line will "separate," then "rejoin." The net result will be momentary cavitation followed by a pressure shock. Avoidance is the best solution and is realized by shortening the suction line or, in an existing installation, slowing the line acceleration by gradually opening the discharge valve.

Quick acting flow control is an operating circumstance, thus the acceleration head associated with it should be accounted for in NPSHA. Davidson [1.5] suggests 1.5 ft (0.5 m) for suction lines whose total length is less than 70 diameters. For longer lines, more detailed analysis is warranted, since the control characteristic and the length of discharge line now have an effect.

6.18. NPSH Margin

For centrifugal pumps, NPSHR is determined by a specific performance drop, thus NPSHA should exceed NPSHR by some margin if the pump is to operate without the performance drop. Further justification for a margin arises from the need to avoid cavitation erosion and to account for any system-induced transient reduction in NPSHA.

Adding margins to margins is to be avoided (see "Understating NPSHA"), so the following practice should be adopted (see Fig. 6.28a):

Specify NPSHA normally available at rated flow after allowing for liquid effects, dissolved gas, and acceleration head, if any of these are applicable.

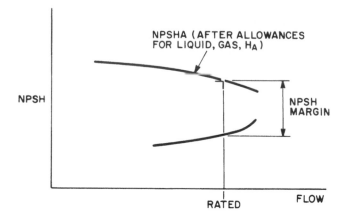

(a) NPSH MARGIN

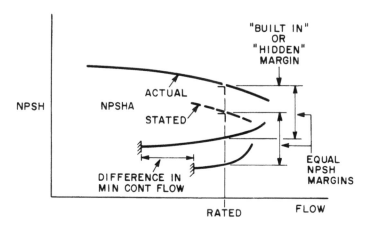

(b) EFFECT OF "BUILT IN" MARGIN

FIG. 6.28. NPSH margin: definition and effect of "hidden" margins.

Specify the required NPSH margin. Normally this is 15% of NPSHR with a minimum of 3 ft (1.0 m), but it may be as much as 50% of NPSHR to account for transients or to avoid localized erosion in high energy pumps.

When a pump must run for extended periods below its suction recirculation capacity, it is possible to avoid noise and impeller erosion with the right combination of S, NPSHR, NPSH margin, and impeller material. The analysis is complicated and only warranted if there is no alternative. A far better approach is to insure that the NPSHA will permit a pump design whose allowable flow range equals that required.

6.19. Understating NPSHA

In Chapter 4 mention was made of the practice of "building in" margins or understating NPSHA to try to insure cavitation-free operation. To see how this can do more harm than good, refer to Fig. 6.28(b). When NPSHA is understated, an impeller of high S is necessary to have NPSHR less than NPSHA on the data sheet. The high S impeller, however, has a higher recirculation capacity, and if run much below BEP will be prone to the noise, vibration, and erosion associated with that phenomenon. The net result of understating NPSHA is thus a very limited pump. It would be better to specify the "normal" NPSHA, recognizing that in the event NPSHA fell to the "minimum," some cavitation might occur. The virtue of doing this is a conventional, rangeable pump that may occasionally be distressed rather than an extreme design that is distressed most of the time.

6.20. Chemical Pump Types

Chemical process pumping is a demanding service. Byrd [6.20] lists the following operating capabilities of chemical process pumps for satisfactory performance in a typical duty:

1. Cope with liquids containing quantities of grit, scale, catalyst, and gas.
2. Withstand occasional suction blockages and running empty or against a closed delivery.
3. Continue running reliably even when impeller and shaft sleeve [have] suffered some corrosion (loss of balance).
4. Withstand normal pipework strains, knocks, and bangs.
5. Have minimal leakage, especially when handling corrosive, valuable, toxic, or flammable liquids.

Of the basic pump types, the centrifugal pump is best suited to realizing the above capabilities. This potential, plus broad hydraulic coverage and availability in most materials, accounts for the extensive use of centrifugal pumps in chemical process service.

A wide variety of centrifugal pump configurations is employed for chemical process service. Figure 6.29 shows those normally used. Not all the configurations shown in Fig. 6.29 possess the capabilities listed above. And of those that do, not all will be of equal capability, there being a distinct connection between ruggedness and cost and efficiency. Increasing ruggedness for a given hydraulic performance will only be realized at the expense of capital and energy.

Figure 6.29 divides the normally used centrifugal pumps into two categories: sealed and sealless. The distinction reflects increasing concern with pollution, a concern that has made the choice of sealed or sealless a basic decision in determining pump configuration.

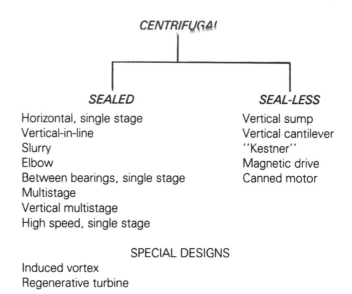

FIG. 6.29. Centrifugal pump types for chemical process.

The various configurations shown in Fig. 6.29 are described in the following text.

6.20.1. Sealed Centrifugal Pumps

6.20.1.1. Horizontal, Single Stage Pumps

By Hydraulic Institute classification, most of the horizontal, single stage pumps used in chemical process service are overhung, separately coupled, and foot or centerline mounted. While the close coupled, casing mounted version has considerable conceptual merit, it is not widely used, difficulty with the motor shaft extension probably being the prime obstacle.

Foot-mounted chemical pumps are built to three standards around the world: ANSI B73.1M-1987 [6.21] in the United States and ISO-2858 [6.22] and DIN-24256 [6.23] in Europe. Pumps built to these standards are of nominal hydraulic performance and specific dimensions, within a particular standard, thus allowing initial plant layout and subsequent pump exchange regardless of pump make.

Standard chemical pumps are of the back-pull-out arrangement to permit dismantling the pump without disturbing the piping or driver (the latter requires a spacer coupling). The shaft seal is usually a mechanical seal, occasionally a packed box. For severe services some manufacturers offer a dynamic seal. (See Chapter 11 for a detailed discussion of shaft seals.) Bearings are antifriction, oil lubricated, natural circulation air cooled; forced air or water cooling options are available from some manufacturers.

The only major differences between standard chemical pumps are impeller closure and impeller attachment. Impellers may be open, semi-open, or closed. Open and semi-open are less prone to clogging and can be adjusted axially to compensate for wear (which requires a cartridge-mounted thrust bearing). Closed impellers afford a simpler pump, lower sensitivity to mechanical and thermal distortion, and are claimed to wear at a lower rate in some circumstances. Impeller attachment may be threaded (rotation to tighten) or keyed. Threaded is simple and resists fretting, but is prone to galling (unless thread clearance is large) and will unscrew if the pump is started in reverse. Keyed offers improved concentricity, immunity to unscrewing and easier dismantling, but is more complex and susceptible to fretting.

Noting general practice, it seems that open and semi-open impellers, with essentially axial running clearances, are usually threaded, while closed impellers, which normally have radial clearances, are keyed.

Figures 6.30 and 6.31 show typical ANSI chemical process pumps, the former having a semi-open impeller and the latter a closed impeller.

For duties beyond the capability of standard chemical pumps (hydraulic

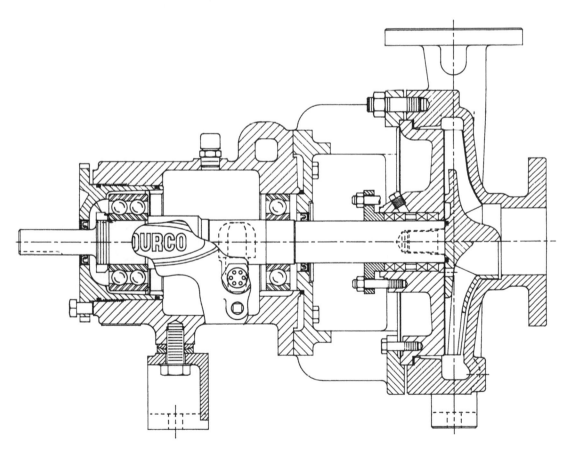

FIG. 6.30. ANSI chemical pump with semi-open impeller. (Courtesy The Duriron Company, Inc.)

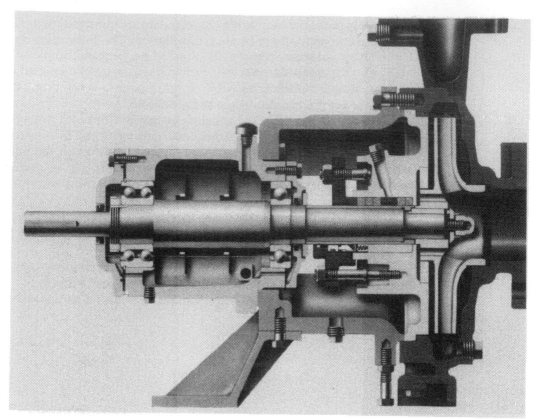

FIG. 6.31.   Closed impeller ANSI chemical pump. (Courtesy Worthington Pump, Dresser Industries, Inc.)

performance, pressure rating, or pumping temperature), the usual choice is refinery process pumps built to API-610 [6.24].

Overhung API-610 pumps are arranged for back-pull-out and are generally centerline supported, a specification requirement above 350°F (175°C). The shaft seal is normally mechanical with a packed box an option, but interchangeability within the same seal housing is not required. The bearings are antifriction (duplex 7300 series thrust bearing), oil lubricated, with cooling necessary to keep oil within 180°F (82°C) or 70°F (39°C) above ambient. The impeller is closed and keyed to the shaft. Pump-out vanes for axial thrust balancing are precluded. Specific pump shaft displacement limits are given for applied nozzle forces and moments, with the option of a simple test to verify pump/base stiffness. Figure 6.32 shows a typical overhung API-610 pump.

There is opinion, e.g., Bloch [6.25], that chemical processing requires a horizontal, overhung pump whose construction lies between ANSI and API. The areas of concern are main joint integrity, seal housing dimensions and versatility, bearing capacity, and bearing lubrication. The 1987 issue of ANSI

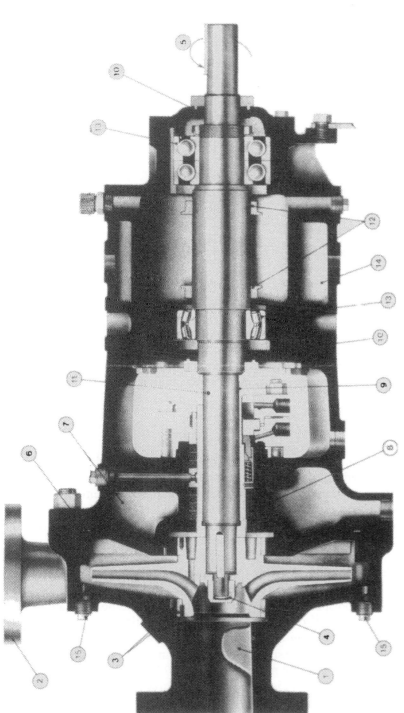

1. Guide vane reduces vortexing and assures uniform flow to the pump inlet.

2. 300 lb. raised face flanges in accordance with ANSI B16.5 standards. 600 lb. flanges are optional.

3. Renewable casing and impeller wear rings are held in place by locking pins or threaded dowels.

4. Impeller acorn nut is threaded to tighten during rotation, preventing leakage and corrosion on the threads.

5. Counterclockwise rotation (CCW) when viewed from the coupling end.

6. Metal to metal fit with confined, controlled compression gasket ensures proper sealing and alignment.

7. Integrally cast stuffing box cooling water jacket has four symmetrical connections for maximum piping flexibility.

8. Seal housing designed to accommodate single, double or tandem mechanical seals. Packed box also available.

9. Replaceable shaft sleeve extends beyond the seal plate for easy leakage detection.

10. Labyrinth-type bearing seals retain oil and keep foreign material out of the housing.

11. Large diameter shaft with short overhang reduces deflection to a minimum.

12. Two oil flingers assure superior lubrication by circulating cool oil through the bearings.

13. Line bearing is double-row spherical roller; thrust bearings are dual, single-row angular-contact ball-type, 7000 series, 40-degree contact angle, installed back-to-back.

14. Annular cooling jacket is cast integrally with the bearing housing.

15. Vent and drain connections are optional.

FIG. 6.32. Overhung, end suction, centreline supported API 610 process pump. (Courtesy Worthington Pump, Dresser Industries, Inc.)

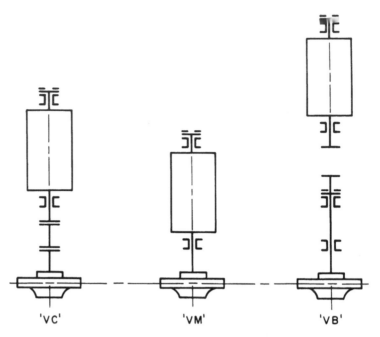

FIG. 6.33. Vertical in-line pump rotor arrangements; ANSI B73.2 designations.

B73.1 addresses one of these concerns. Byrd [6.20] reports experience supporting the need for versatility in seal housing arrangement (see Ch. 11, seals). Manufacturers are offering options of some or all of the desired changes.

6.20.1.2. Vertical In-Line Pumps

An in-line pump, one whose nozzles are the same size and in-line or coaxial, offers simplified installation. Making the shaft axis vertical avoids the effect of gravity on rotor deflection and reduces floor area requirements. Connecting the driver directly to the pump, for which Anderson [6.26] makes a cogent argument, eliminates misalignment caused by piping forces. Combining these features yields the vertical in-line pump.

For chemical process service, vertical in-line pumps are built to two standards: ANSI B73.2M-1984 [6.27] in the United States and BS4082-1966 [6.28] in Europe. ANSI B73.2M-1984 stipulates nominal hydraulic performance and specific dimensions, while BS4082-1966 stipulates dimensions only. BS4082 also covers a second configuration, designated "U," in which the nozzles are on one side of the pump and parallel to each other.

Within the general classification "vertical in-line" pumps, three distinct rotor arrangements are employed. ANSI B73.2M-1987 designates these "VC," "VM," and "VB"; Fig. 6.33 shows the three arrangements.

VC. The impeller and seal are mounted on a stub pump shaft which is connected to the motor shaft with a rigid, spacer coupling. This arrangement allows removal of the pump's rotor and seal without disturbing the motor. The inclusion of a coupling increases the impeller overhang and makes rotor concentricity harder to maintain, but with appropriate liquid end design and careful manufacture, these need not be deficiencies.

VM. The impeller and seal are mounted directly on the motor shaft. Pumps built to ANSI B73.2 use motors with an extended shaft to NEMA dimensions. Some manufacturers build pumps to API-610 by using a standard motor shaft with a shrink fit stub extension. In either version, dismantling the pump requires removal of the motor.

VB. The pump impeller, seal, and shaft are supported by a separate bearing frame. A flexible spacer coupling connects the pump and driver, allowing

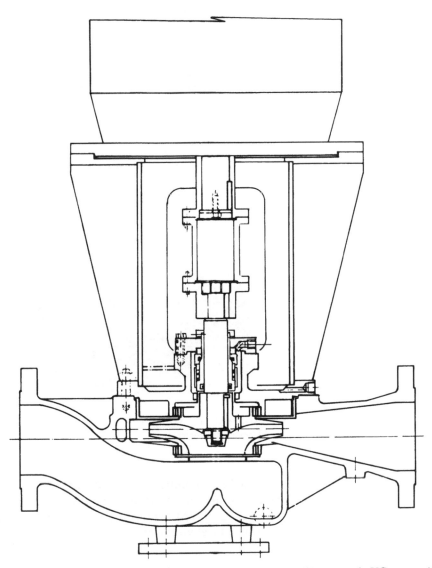

FIG. 6.34. Rigidly coupled vertical in-line process pump; ANSI rotor-style VC; meets API 610. (Courtesy Worthington Pump, Dresser Industries, Inc.)

FIG. 6.35.   ANSI-style VM vertical in-line chemical pump. (Courtesy Worthington Pump, Dresser Industries, Inc.)

the pump to be dismantled without disturbing the driver. Because there is a separate pump-bearing frame, the driver's center of gravity is higher than in the VC or VM arrangements, thus more care is needed to avoid resonant structural vibration.

Shaft seal options are the same as those for horizontal pumps. Bearings are antifriction, grease lubricated, natural or forced circulation air cooled. "VC" and "VM" arrangements require the motor bearings to accommodate

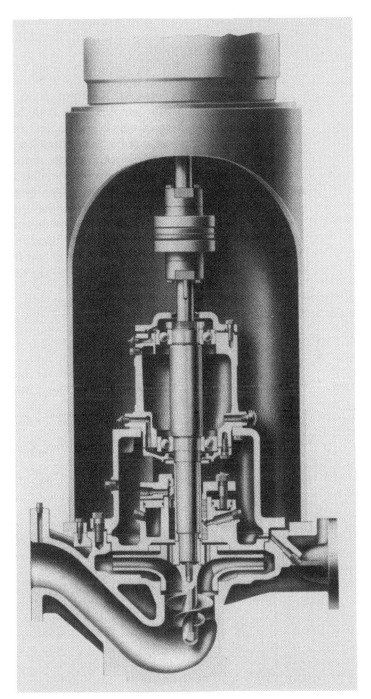

FIG. 6.36. Separately coupled vertical in-line chemical pump; ANSI-style VB. (Courtesy Worthington Pump, Dresser Industries, Inc.)

both pump and motor rotor loads; for high temperature service, various thermal barriers are available to reduce heat conduction to the motor.

Figures 6.34, 6.35, and 6.36 show various offerings for the three arrangements. Which to choose depends upon past experience, particular service, plant maintenance practices, and personal preference.

6.20.1.3. Slurry Pumps

Slurry pumping is usually associated with mining and mineral processing. In chemical processing, however, there are services where the combined effects of erosion and corrosion (see Section 3.2) are better handled by a significant increase in erosion resistance. As a general rule, slurry pumps should be considered whenever the solids concentration exceeds 2–3%. In some services, however, hard, sharp particles at concentrations as low as 1% can warrant slurry pumps. Such a service is scrubber liquid circulation in flue gas desulfurization, for which rubber-lined slurry pumps, with corrosion-resistant treatment of the shaft seals, are now the usual selection.

Conventional slurry pumps have an overhung rotor, are single stage, and generally end suction (there are side suction versions, the objective being to have the shaft seal at suction pressure). Two forms of liquid end are used: rubber lined and hard metal. Casings for the former are radially split with a

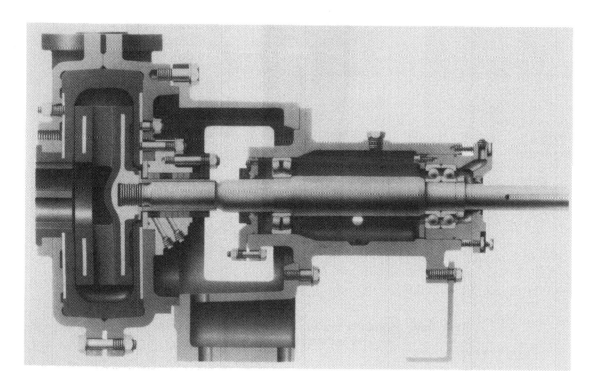

FIG. 6.37.   Rubber-lined slurry pump (Worthington Pump, Dresser Industries, Inc.)

rubber liner in each half, see Fig. 6.37. Those for hard metal pumps may be "solid" with bolt in wear plates, see Fig. 6.38, or radially split with a "solid" hard metal liner, see Fig. 6.39. Impellers are usually closed (more even wear and greater tolerance of wear), with pump-out vanes back and front. Rubber impellers are molded onto a metal skeleton. Impeller attachment is usually threaded, there being a need to avoid joints in the flow path (each discontinuity is a potential erosion site).

Shaft sealing has traditionally been packed box or dynamic. Recent progress with mechanical seals, however, has yielded designs of promise. The virtue of these most recent designs is the elimination of seal flush, hence an auxiliary system and product dilution, or avoiding the limitations of dynamic seals; see Chapter 11 for further discussion.

An essential feature of slurry pump design is tolerance of severe wear. On top of that, the pumps usually operate in a rugged environment with only rudimentary maintenance. For rotor construction these requirements dictate

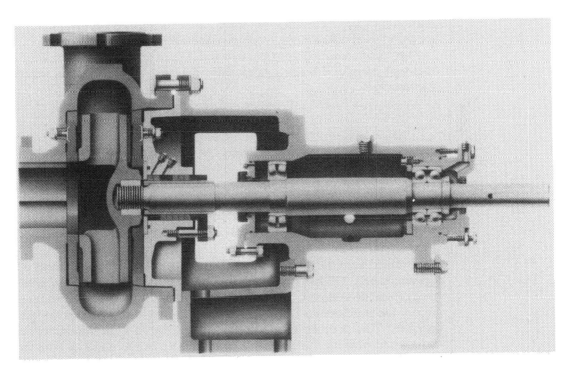

FIG. 6.38. Hard metal slurry pump with replaceable wear plates. (Courtesy Worthington Pump, Dresser Industries, Inc.)

FIG. 6.39.   Hard metal slurry pump with a "solid" casing liner. (Reprinted with permission of Warman International, Inc.)

high shaft stiffness, high bearing capacity (L_{10} life is typically 50,000 h at maximum speed), bearings well sealed from the environment, and a simple means of rotor adjustment to compensate for liquid end wear.

The head a slurry pump can develop is limited by two factors: wear and material strength. Precise relationships for wear are elusive, but a general expression is

$$\text{wear rate } \alpha \text{ (velocity)}^m$$

or, since in a centrifugal pump head is proportional to velocity squared

$$\text{wear rate } \alpha \text{ (head)}^{m/2}$$

where the exponent m ranges from 2.4 to 4.0 and depends upon service conditions. From this it is evident only so much head can be developed before the wear rate is unacceptable.

Limited material strength applies to rubber. At peripheral speeds above approximately 100 ft/s, the rubber lining of impellers is prone to breaking away from the skeleton. To try to circumvent this limitation, some manufacturers use hard metal impellers in rubber-lined casings when the service permits. This practice relies, in part, upon the fact that liquid velocities relative to working surfaces are usually lower in the impeller than in the casing.

When the system head is greater than can be developed in one pump, series pumping is resorted to. Casing pressure capability and shaft seal pressure need care in such cases.

General guidelines for application are rubber-lined pumps for liquids containing fine solids, hard metal for coarse solids. Chemical process pumping will therefore usually employ rubber-lined slurry pumps.

The hydraulic design of slurry pumps is influenced significantly by the need to resist and tolerate wear. Wear resistance limits the usable range of specific speeds and dictates compromise in the form of some parts. Tolerance of wear further compromises form by dictating greater than usual thickness. The net result is efficiency lower than equivalent pumps designed for clean liquids.

Pumping solids-laden liquid reduces the efficiency of slurry pumps still further. Two factors are cited in explanation: slurries of fine particles display a viscosity effect and velocity energy imparted to solids is not converted to pressure (potential) energy when the mixture is slowed down (diffused). Corrections for slurry performance are controversial, thus manufacturers should be consulted for specific recommendations.

6.20.1.4. Elbow Pumps

Circulating duties involve high flows at low heads, the head being only the friction loss around the circuit. The appropriate pump type for such hydraulic conditions is the high specific speed (see specific speed) axial flow or propeller pump. In chemical process service, propeller pumps are used in the so-called elbow configuration, see Fig. 6.40. The virtue of this configuration is simplified installation.

Elbow pump rotor construction is overhung, with either one internal and one external bearing (older designs) or two external bearings (newer designs) as in Fig. 6.40. External bearings are generally antifriction, oil, or grease lubricated. Shaft seal will be determined by the nature of the pumped liquid, mechanical being the usual choice.

When an internal bearing is used, the pump's suction piping, or a spool in it, has to be removed before dismantling the pump. By incorporating a radial joint in the back of the elbow, designs with two external bearings become back-pull-out pumps, enabling the entire rotor and bearing assembly to be removed and serviced as a unit.

6.20.1.5. Between Bearings, Single Stage Pumps

As their design flow increases, pumps of low to medium specific speed reach a flow where double suction impellers are necessary to yield tolerable NPSHR. Except in unusual designs, double suction impellers are mounted on between-bearings rotors. Further, there is a point where a between-bearings (simply supported) rotor is more economical than an overhung (cantilever) rotor for the same integrity.

FIG. 6.40.   Elbow-type circulating pump. (Courtesy Worthington Pump, Dresser Industries, Inc.)

Compared to overhung end suction pumps, the casings of a single stage between bearings pumps are more difficult to cast in corrosion-resistant alloys (long, thin sections) and use relatively more metal. Compounding this, most between-bearings pumps require two shaft seals versus one for overhung pumps. The net result is that for chemical process service, between-bearings pumps are used only when careful evaluation shows their higher cost to be justified. Generally this will be in larger plants with their inherently higher flows.

Single stage, between-bearings pumps are furnished in two basic casing configurations: axially split and radially split. Figure 6.41 shows a typical axially split pump; Fig. 6.42 shows a radially split pump. Other than the casing joint and casing wearing ring retention, the construction is essentially the same. Shaft seals are mechanical, cartridge mounted. Bearings are usually antifriction, but may be hydrodynamic if load and speed necessitate it. Pumps designed for severe service often incorporate a high capacity thrust bearing, duplex angular contact or tilting pad to raise their tolerance of off-design operation.

6.20.1.6. Multistage Pumps

When the hydraulic duty of a centrifugal pump is such that it cannot be met at regular driver speeds with good efficiency (see specific speed), the traditional solution is to employ a multistage pump.

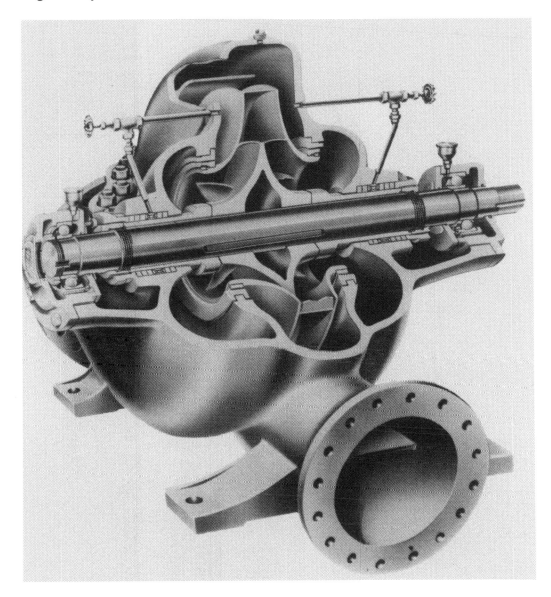

FIG. 6.41. Single stage, axially split, double suction volute pump. (Courtesy Worthington Pump, Dresser Industries, Inc.)

In essence, a multistage pump is a series of single stages, all but the last stage complete with a "return channel" to guide liquid to the succeeding stage, arranged within a single casing. While there are many forms of multistage pumps, the basic distinctions are between shaft orientation, rotor arrangement (impeller mounting and axial thrust balance), and casing construction. Table 6.1 outlines the options and their usual combinations.

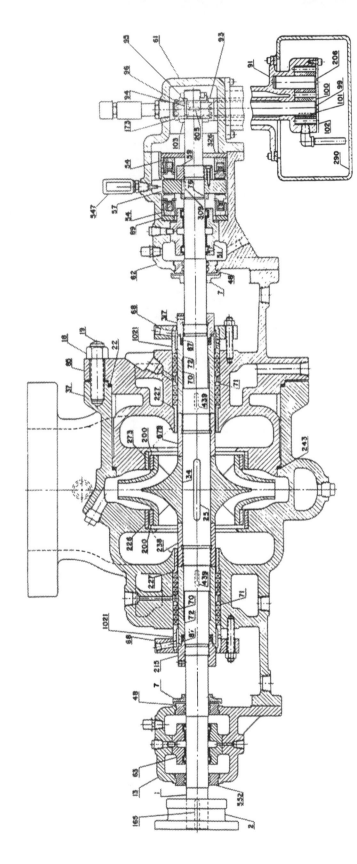

FIG. 6.42. Single stage, radially split, double suction volute pump. (Courtesy Pacific Pump, Dresser Industries, Inc.)

TABLE 6.1 Multistage Pump Configurations[a]

Shaft Orientation:		Horizontal			Vertical	
Casing Construction:		Segment	Axial Split	Barrel	Segment	Can
Impellers	Thrust Balance					
Tandem	None	—	—	—	—	U
Tandem	Individual	U	—	—	—	A
Tandem	Balancing device	U	A	U	U	—
Opposed	Opposed Impellers	A	U	A	—	—

[a]U = usual, A = available.

For chemical process service, the usual casing configurations are axially split, barrel, and can (in modern parlance, both barrel and can are known as "double casing"). Segmental casings are rarely used because of the multiple leakage points inherent in their construction, see Fig. 6.43.

Axially split casings are the first choice for horizontal pumps. They are limited to working pressures of approximately 2000 lb/in.^{2}gauge (140 kg/cm^2), temperatures to 600°F (315°C), and liquids whose SG = 0.7 or greater. The last two limitations are recommended by API-610 and may be exceeded if there is prior satisfactory experience with the casing in question. Designs for working pressures greater than 2000 lb/in.^{2}gauge are feasible but economically dubious.

Barrel casings make for a more complicated pump but are necessary for service conditions beyond the capability of axially split casings.

Most axially split multistage pumps employ opposed impeller rotors. Figure 6.44 shows a typical pump. The claimed virtue of opposed impellers is inherent zero thrust with no hydraulic balancing device to wear. The usual opposed impeller arrangement, as shown in Fig. 6.44, does have some

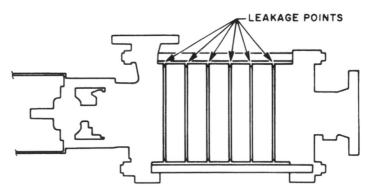

FIG. 6.43. Multiple potential leakage points in a segmental, radially split pump casing.

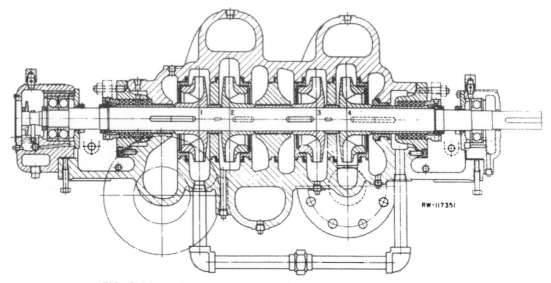

RW-117351

FIG. 6.44. Axially split multistage pump with opposed impellers for hydraulic axial thrust balance. (Courtesy Worthington Pump, Dresser Industries, Inc.)

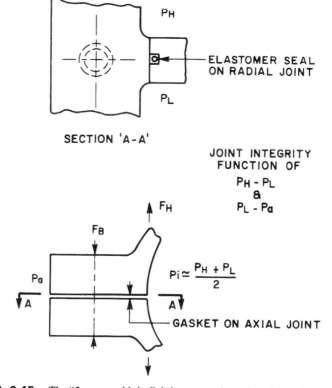

FIG. 6.45. The "3-cornered joint"; inherent weakness in axially split casings.

residual thrust (the product of internal leakage across the center and end bushings) and its magnitude increases as the clearances increase. Despite this, pumps with correctly sized thrust bearings are entirely serviceable. In units of many stages, say six or more, operation of the relatively slender rotors is doubtless enhanced by the Lomakin effect at the center bushing. The Lomakin effect is hydraulic stiffness produced by differential pressure across an annular clearance.

One of the major difficulties in axially split pumps is sealing the inevitable "3-cornered joints," see Fig. 6.45. As the pressure drop across such joints increases, so does the difficulty of sealing them. For this reason most barrel or double casing pumps, whose sole justification is better pressure containment capability, employ radially split elements or inner casings. Opposed impeller rotors are not easily incorporated in radially split elements (the necessary crossovers pose a major problem), thus most barrel pumps employ tandem impeller rotors. Figure 6.46 shows such a pump.

In the interests of simplicity and keeping the bearing span short, all but the smallest tandem impeller rotors employ a balancing device to counteract impeller axial thrust. Three types are used: disc, drum, and stepped drum or combined disc/drum.

The classical balancing disc, Fig. 6.47(a), is entirely self-compensating, automatically varying its axial gap to develop a balancing force equal to impeller thrust. To be effective, the disc's axial gap has to be very small. While this lowers the balancing leak-off flow, thus adding 1 or 2 points to the pump's

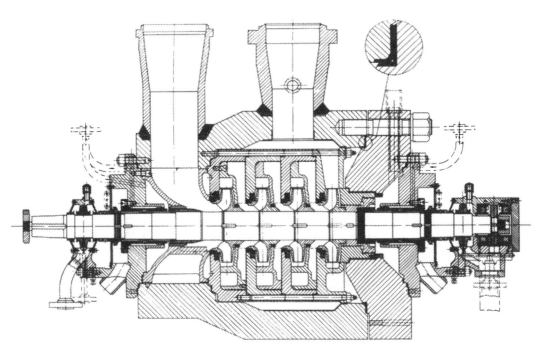

FIG. 6.46. Double casing (barrel) multistage pump with radially split inner element. (Courtesy Worthington Pump, Dresser Industries, Inc.)

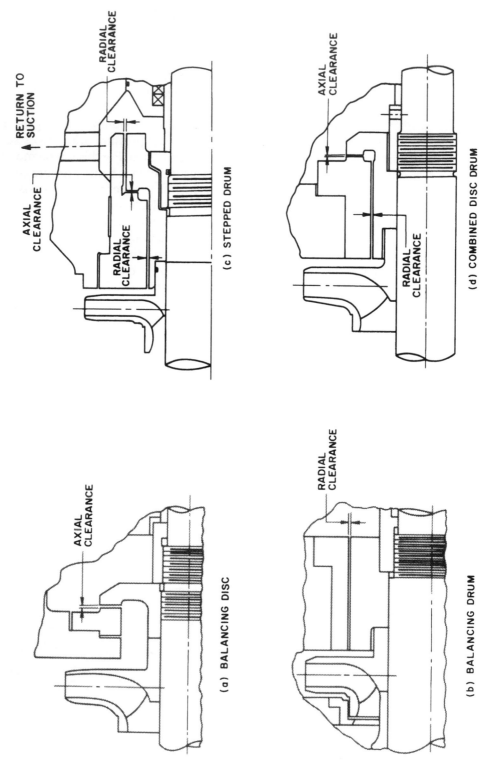

FIG. 6.47. Devices for balancing hydraulic axial thrust in multistage pumps.

efficiency, the close axial clearance is prone to incidental contact during operating transients. As a consequence, balancing discs are not widely used in barrel pumps.

By resorting to a balancing drum (often called a piston although it does not reciprocate), Fig. 6.47(b), the rotor's sensitivity to close axial positioning is eliminated. The drum, however, is not self-compensating, meaning there will be residual axial thrust over some or all of the pump's operating flow range. To avoid rotor "shuttle," some inboard residual thrust is desirable and balancing drums are usually so sized. Changing the thrust balance of a drum requires that its diameter be changed.

If a balancing drum is modified to incorporate an axial clearance, it acquires a degree of self-compensation, the degree depending upon the geometry of the axial clearance. Two arrangements are employed to incorporate the axial clearance: a stepped drum, Fig. 6.97(c), and a combined disc/drum, Fig. 6.47(d). With its "disc" component between the two major pressure breakdowns (radial clearances), a stepped drum allows "fine tuning" thrust balance by changing either the minor or major diameter clearance. To avoid incidental contact at the axial clearance, most stepped drums or combined disc/drums are set such that their self-compensating capability will only come into operation if the normal means of rotor positioning fail.

Rotor design for multistage pumps is a controversial topic. The basic argument centers on the relationship between the mechanical stiffness of the rotor and the hydraulic stiffness (Lomakin effect) generated in the clearances surrounding it. One school, citing lower cost and higher efficiency as justification, argues for low mechanical stiffness (slender shaft) and high hydraulic stiffness. The other school, arguing that reliability outweighs first cost and efficiency considerations, promotes high mechanical stiffness (large shaft) and low hydraulic stiffness. It is interesting to note that the argument is peculiar to liquid-handling turbomachines; those handling gases inherently have only minimal hydraulic stiffness, thus mechanical stiffness has to be high.

Figure 6.48 shows qualitatively the effect of the two approaches. Rotors of low mechanical, high hydraulic stiffness have a low "dry" critical speed but are classically "stiff" (first bending critical above the running speed) at 100% speed. Those of high mechanical, low hydraulic stiffness have a higher "dry" critical speed but are little affected by running speed, being classically "flexible" at 100% speed.

Hydraulic stiffness is dependent upon the form and magnitude of running clearances. For high hydraulic stiffness, running clearances must be close and of smooth surfaces. As the clearances increase, hydraulic stiffness diminishes, lowering the rotor's critical speed. As shown in Fig. 6.48, the reduction in hydraulic stiffness can be enough to bring the critical speed into coincidence with the running speed. If this occurs when the clearances offer little damping, destructive rotor vibration can set in. Because their critical speed is always below the running speed, for constant speed pumps anyway, rotors of low hydraulic stiffness are immune to this particular problem.

Coincidence between critical and running speeds is not the only form of destructive vibration to multistage pumps. Subsynchronous vibration, par-

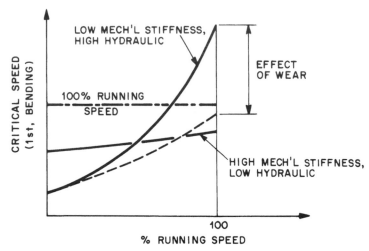

FIG. 6.48. Comparison of rotor critical speeds against running speed for two design philoso-
phies employed in multistage pumps.

ticularly in pumps of high energy density (energy absorbed/rotor weight), is a
vexing problem. Whether this is caused by classical instability (self-excited
vibration) or resonance with subsychronous excitation is still a moot point,
and one beyond the scope of this volume.

There is no clear choice between the two rotor design approaches. As a
start, the following points need to be considered.

Neither design approach is tolerant of high hydraulic forces, so if the hydraulic design is poor or the selection poor or both, pump life will be short.

For continuous service at constant flow pumping benign liquids, slender shaft
rotors have realized service lives of the order of 50,000 h. Under similar
conditions, large shaft rotors have run in excess of 150,000 h between
overhauls.

When the service is start/stop over a wide flow range or pumping a liquid
containing some solids or of low lubricity, slender shaft rotors are susceptible to rapid wear at their clearances. Employing materials (e.g., carbon,
Teflon, ceramics) to turn some of the clearances into internal bearings is a
potential solution, but so far has only proven viable in small pumps of low
energy density. Larger pumps with slender shaft rotors may have their
service lives reduced to only 7,500–10,000 h, those with large shaft rotors
to 30,000–40,000 h. In really severe services, wear rate can preclude the
use of slender shaft rotors and reduce the service life of large shaft rotors to
7,500–10,000 h.

Duncan and Hood [6.29] provided a quantitative guide for rotor mechanical
stiffness. By using their chart, repeated in Fig. 6.49, it is possible to classify
a particular rotor for a given speed as "too slender," "slender," capable of
"wet running" only, or good for "dry running."

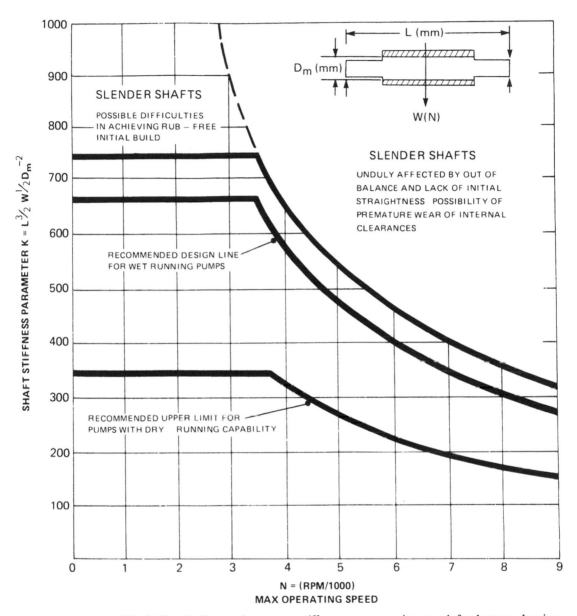

FIG. 6.49. Guidance chart; rotor stiffness versus running speed for between bearings rotors [6.29].

$$K = \left(\frac{WL^3}{D_m^4}\right)^{1/2} \qquad \text{N, mn units}$$

$$= 3.132\left(\frac{WL^3}{D_m^4}\right)^{1/2} \qquad \text{kg, mn units}$$

$$= 0.418\left(\frac{WL^3}{D_m^4}\right)^{1/2} \qquad \text{lb, inch units}$$

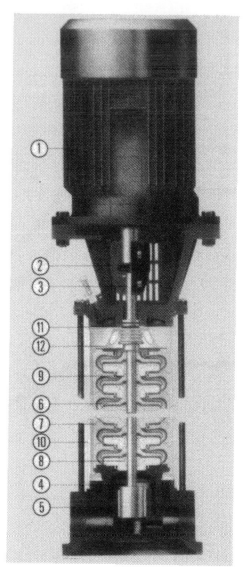

FIG. 6.50. "Standard" vertical multistage pump. (Courtesy Worthington Pump, Dresser Industries, Inc.)

Shaft seals are usually cartridge-mounted mechanical. Packed box seals are optional but are used only for innocuous services or when a mechanical seal is deemed too complicated. With either seal type, but particularly with mechanical, it is advisable to ensure the inboard (coupling end) seal can be replaced without opening the casing or moving the driver. A spacer-type coupling with the pump hub taper mounted and a seal flange that will pass through the bearing bracket fit are the principal features needed to ensure this capability. The taper-mounted hub, while more expensive, permits easy removal in a confined space.

Bearings can be either antifriction or hydrodynamic. Antifriction bearings are limited to smaller pumps. Typical antifriction bearing types are a single row deep groove ball or self-aligning ball for the line bearing and a duplex angular contact for the thrust bearing.

Hydrodynamic bearings are employed when the speed or load preclude antifriction bearings. Journal bearings are usually plain or profiled bore sleeve, there being little justification for tilting pad journal bearings in pumps. Thrust bearings are almost exclusively tilting pad, generally in a self-leveling mounting. To avoid the heat load generated by tilting pad thrust bearings, some pump designs, i.e., those whose axial thrust is fairly predictable, employ a small antifriction thrust bearing in conjunction with hydrodynamic journal bearings. Oil lubrication is predominant, though some hydrodynamic bearings in special designs are lubricated by the pumped liquid.

6.20.1.7. Vertical Multistage Pumps

Vertical multistage pumps have, of necessity, pumped liquid lubricated bearings throughout their liquid end. Because the bearing/lubrication arrangement is less than ideal, vertical multistage pumps generally have shorter service lives than their horizontal equivalents. For this reason they are usually used only when the application dictates it. The most common application dictate is low NPSHA, thus most vertical multistage pumps in chemical processing are of the can type. With this configuration the NPSHR is provided by lowering the first stage impeller.

Can-type pumps employ a segmental casing, but as the casing is contained within the can, the risk of a high pressure leak to atmosphere is greatly reduced. Such is not the case with mass produced vertical segmental pumps of the type shown in Fig. 6.50. This, plus limited material options, a consequence of mass production, restricts the services in which such pumps can be used.

Axial thrust in can pumps can be either balanced by back-wearing rings on each impeller, see Fig. 6.51, or unbalanced and applied to a suitable thrust bearing, see Fig. 6.52. Balanced construction becomes necessary as pump total head and rotative speed increase.

The usual vertical multistage pump, the can pump, has discharge pressure at its shaft seal unless a breakdown bushing is used. For this reason the shaft seal is generally mechanical, with cartridge mounting for ease of installation. Recognizing that most can pumps are used to pump liquids near their boiling point (low NPSHA), it is evident that having discharge pressure at the seal inherently provides the margin over vapor pressure, at least 25 lb/in.^{2}, necessary to avoid flashing at the seal. Note, however, that the pressure at the seal of a standby pump is suction pressure. For some applications this dictates special arrangements to prevent "air in" leakage. Liquid end and line shaft bearings are, as noted earlier, sleeve type, product lubricated. The thrust bearing may be antifriction or hydrodynamic depending upon load and speed. In United States practice the thrust bearing is normally incorporated in the motor, while European practice is to employ a separate thrust bearing.

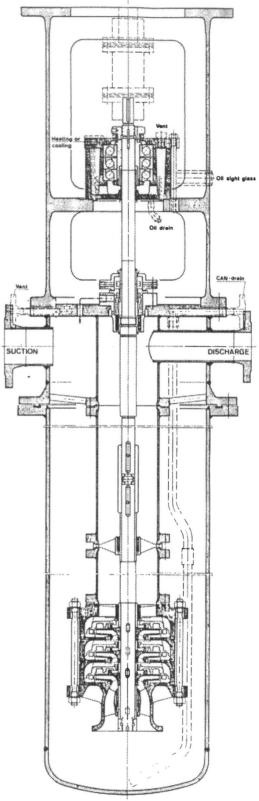

FIG. 6.51. Vertical, double casing (can) pump with individually balanced impellers; residual hydraulic axial thrust and rotor weight supported by separate thrust bearing. (Courtesy Worthington Pump, Dresser Industries, Inc.)

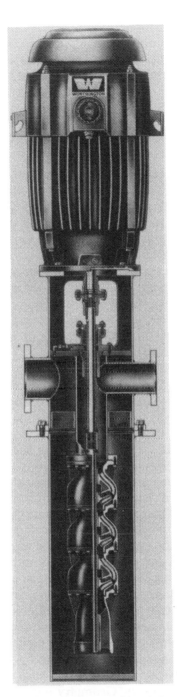

FIG. 6.52. Vertical, double casing (can) pump with unbalanced impellers; hydraulic axial thrust and rotor weight supported by motor thrust bearing. (Courtesy Worthington Pump, Dresser Industries, Inc.)

For the former, the pump lineshaft is rigidly coupled to the motor, with provision for axial adjustment. With a separate thrust bearing, a flexible coupling is employed. For either arrangement the coupling is a spacer type to permit seal replacement without having to disturb the driver.

6.20.1.8. High Speed, Single Stage Pumps

As capacity decreases, physical size and the number of stages often required make it difficult to design multistage pumps with high rotor mechanical stiffness. Such designs are thus entirely dependent upon hydraulic stiffness for satisfactory operation. When the liquid pumped has low SG, viscosity, and lubricity, pumps dependent upon hydraulic stiffness for rotor support tend to be unreliable.

An alternative approach for many low flow, high head duties is the single stage, high speed pump. In these designs the pump's rotative speed is high enough to develop the required head in a single stage. The pump selection chart in Fig. 6.13 identifies coverage (Region 3) for these pumps with NPSHR limited to 30 ft. Current practice has single stage pumps running up to 25,000 r/min and developing heads of 7000 ft (heads to 12,000 ft with up to three pumps in series).

High speed, single stage pumps offer high rotor integrity, low weight, simple construction easy to produce in corrosion resistant materials, and relatively straightforward maintenance (provided allowance is made for the needs of a precision, high speed machine). The drawbacks are higher NPSHR, the need for a high speed drive, and no tolerance of solids in the pumped liquid (see slurry pumps, head versus wear rate).

Two forms of high speed, single stage pumps are in use today. The original version, described by Barske and cited by Lobanoff and Ross [6.30], is known as a "partial emission" pump. The term "partial emission" refers to a hydraulic design where the impeller discharge area is significantly greater than the casing throat area. Pumps following the Barske design have radial vaned, open impellers. This configuration is relatively insensitive to impeller to casing clearance, hence is not generally affected by wear. Standard designs have a head characteristic which is flat or slightly drooping below BEP and very steep beyond BEP, see Fig. 6.53. Impellers of high solidity (more vanes) have a constantly rising head characteristic and produce less noise. Figure 6.54 shows a typical Barske pump.

The second version employs a more conventional hydraulic design to achieve a constantly rising head characteristic, with sufficient head rise for controllability. Geometry (eye diameter to outside diameter) puts a lower limit on the flow of such designs, see Fig. 6.13. Internal leakage is limited by conventional wearing rings, thus making these pumps less tolerant of wear than the Barske design. A typical high head, single stage pump of conventional hydraulic design is shown in Fig. 6.55.

Both versions are furnished with inducers when it is necessary to reduce NPSHR.

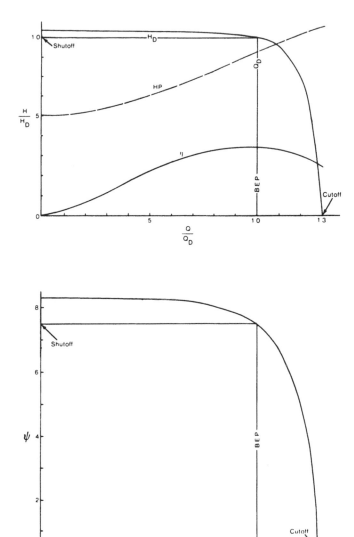

FIG. 6.53. Typical head capacity characteristic of Barske partial emission pump. (Courtesy Sunstrand Fluid Handling, a unit of Sunstrand Corp.)

Shaft seals are exclusively mechanical, of the stationary compression unit arrangement due to the very high nominal seal surface speed. Bearings are hydrodynamic, again a consequence of the high rotative speed. Commercially available units have motor gear drive with the impeller mounted directly on the extended pinion, thus reducing the number of high speed bearings and eliminating a high speed coupling.

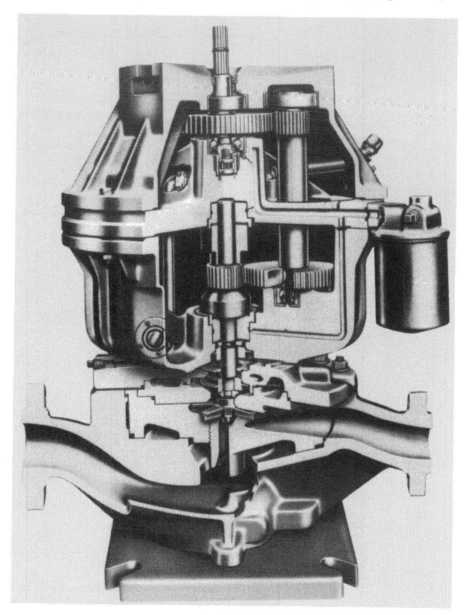

FIG. 6.54. High speed, single stage pump, Barske type. (Courtesy Sunstrand Fluid Handling, a unit of Sunstrand Corp.)

6.20.2. Sealless Pumps

Conventional centrifugal pumps employ a seal to isolate the pumped liquid from the atmosphere where the shaft passes through the casing. Chapter 11 details the various seal types.

In many applications, service conditions or the nature of the liquid pumped make sealing so difficult or zero leakage so important that it becomes necessary to resort to some form of sealless pump.

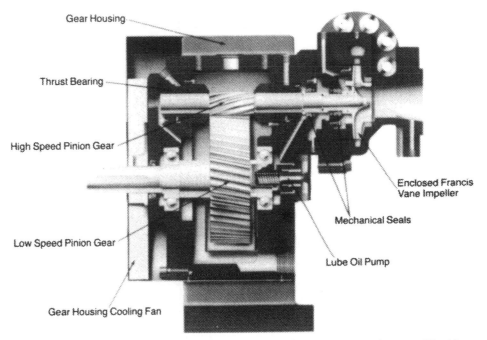

Gear Housing

Thrust Bearing

High Speed Pinion Gear

Low Speed Pinion Gear

Gear Housing Cooling Fan

Enclosed Francis
Vane Impeller

Mechanical Seals

Lube Oil Pump

FIG. 6.55. High speed, single stage pump, Francis-type impeller. (Courtesy Worthington Pump, Dresser Industries, Inc.)

Sealless pumps can be divided into two classes: Pumps with a shaft passing through the casing but employing a configuration, which avoids the need for a liquid seal; and pumps which are hermetically sealed, that is, do not have a shaft passing through the pressure boundary.

6.20.2.1. Pumps Without a Liquid Seal

For chemical process service, four configurations are in common use and a fifth is of note. They are: vertical suspended or sump pump, vertical cantilever pump, the "Kestner" pump, a version of the vertical turbine pump, and the vertical volute pump. Figures 6.56 through 6.60 show the five arrangements in the order listed. The first two have their liquid end in the pumped liquid (wet pit), the third has its liquid end outside the suction vessel (dry pit), and the last two are traditional deep setting wet pit pumps.

All five of this first class of sealless pump rely upon liquid level in the suction vessel and pressure in the suction vessel close to atmospheric to avoid a liquid seal. Two of the designs, Figs. 6.58 and 6.59, also require a breakdown bushing with bleed-off back to suction.

If the pumped liquid vapor must also be isolated from the atmosphere, then a vapor seal is necessary. In most cases this can be a simple grease-filled bulkhead-type seal such as is shown in Fig. 6.61.

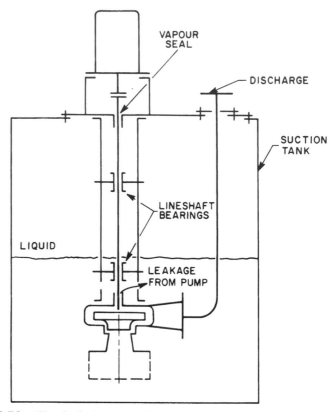

FIG. 6.56. Wet pit, single stage sealless pump; bearing below the sump liquid level.

Given the need for a vapor seal, there is a temptation to not classify the above pump configurations as sealless. The question can be argued either way, but there is no escaping that these configurations have provided reliable service where conventionally sealed pumps could not.

Vertical suspended (sump) pumps are dependent upon their line bearings for reliable operation. Bearing design (dimensions and materials) and lubrication are thus critical. Unless the pumped liquid is clean or can be cleaned, the process must tolerate dilution by an external lubricant. When all these factors are taken into account, sump pumps are usually as expensive as alternative solutions. Low cost sump pumps are generally suitable for intermittent service only.

Vertical cantilever pumps do not have any form of rotor support within or close to the pumped liquid, thus avoiding the limitations of vertical sump pumps. Provided the rotor design is correct, vertical cantilever pumps are tolerant of a wide range of liquids, being limited only by liquid end configuration and materials. They cannot easily be constructed for deeper settings, thus suction vessel shape and level control need more attention than with vertical sump pumps.

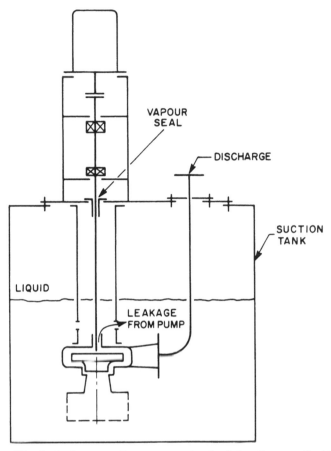

FIG. 6.57. Wet pit, single stage sealless pump; no bearing below the sump liquid level.

The "Kestner" pump is a dry pit version of the vertical cantilever pump. Being so, it requires an adequately sized bleed-off or overflow to return leakage from the pump back to the suction vessel.

Vertical turbine pumps of the configuration shown in Fig. 6.59 rely upon a close clearance breakdown bushing to reduce pump discharge pressure to suction. To function, breakdown bushings must have a flow through them and the flow must be bled off with a minimal pressure drop. The leakage flow represents a direct energy loss, thus every effort is made to minimize it. For this reason, any pump employing breakdown bushings is limited to clean liquids; the rapid erosion associated with abrasive liquids would impose a high cost in power consumption and maintenance.

Vertical volute pumps with enclosed lineshafts, Fig. 6.60, have the inner end of each of the impeller bearings subjected to impeller suction pressure. The advantages of this arrangement are no liquid seal and ease of excluding pumped liquid from the impeller and lineshaft bearings, if that's necessary. Ranked in tolerance of solids laden liquid, the vertical volute pump is second

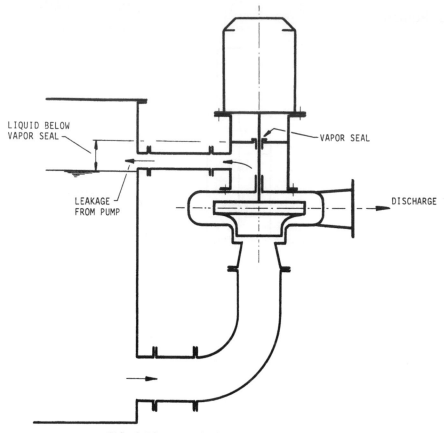

FIG. 6.58. Dry pit, single stage sealless pump.

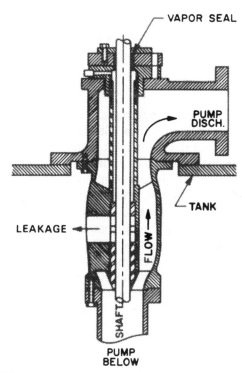

FIG. 6.59. Sealless arrangement for multistage wet pit pump.

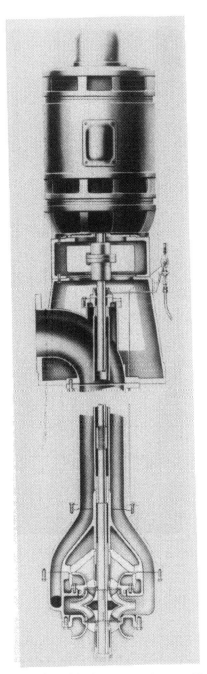

FIG. 6.60.   Sealless configuration of single stage wet pit pump. (Courtesy Worthington Pump, Dresser Industries, Inc.)

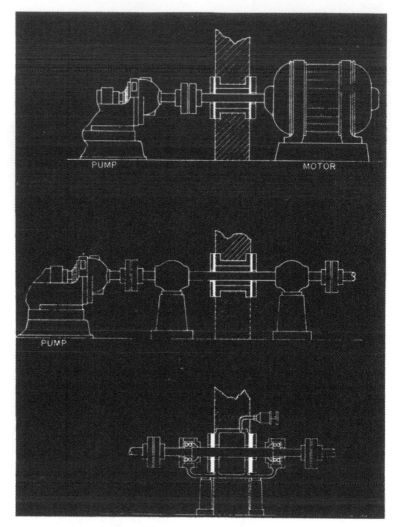

FIG. 6.61. Bulkhead type seals.

to the vertical cantilever. To realize this capability, the process usually has to be tolerant of some dilution by a clean liquid to lubricate and cool the impeller and lineshaft bearings. If the pumped liquid can be easily cleaned to lubricating quality, dilution can be avoided.

Pumps employing dynamic seals (expellers, see Chapter 11) are sometimes classified as sealless. This classification is valid in that the running pump does not have a liquid seal at the shaft opening, but it overlooks the need for some form of liquid seal when the pump is shut down, and for that reason this text classifies such pumps as sealed.

6.20.2.2. Hermetically Sealed Pumps

For applications where the suction pressure or required pump configuration preclude the use of pumps with vapor seals (see above), it is necessary to resort to the second class of sealless pumps, those whose shaft does not pass through the pressure boundary. Such pumps are also referred to as hermetically sealed.

With the limitation of no shaft through the pressure boundary, the first problem in hermetically sealed pumps is how to drive the pump rotor. Two general solutions are used: a magnetic coupling to transmit torque through the pressure boundary and a close coupled electric motor within the pressure boundary.

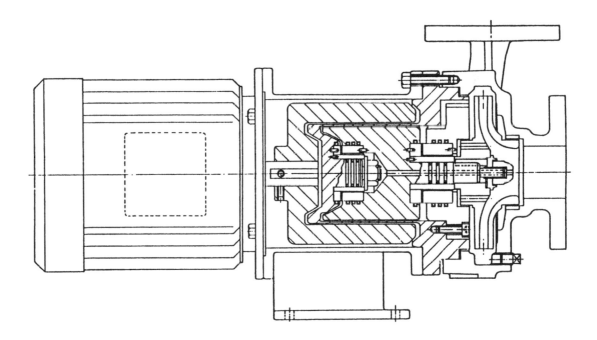

Sectional Arrangement of 'WRM' Pump
(Silicon Carbide Bearings)

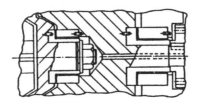

Scrap Section Showing Alternative

FIG. 6.62. Hermetically sealed pump, magnetic drive type. (Courtesy Worthington Pump, Dresser Industries, Inc.)

Magnetic Drives.   Figure 6.62 shows a magnetic coupling driven centrifugal
pump. An outer magnet ring, supported by a separate bearing frame or the
driver rotor, passes a rotating magnetic flux through the containment can
(pressure boundary) to cause rotation of the pump rotor. Two drive principles
are used: eddy current and synchronous.

In eddy current drives, Fig. 6.63, inner ring magnets are created by
induced eddy currents, the product of relative motion (slip) between the
coupling rings. Slip increases with the torque transmitted. Synchronous
drives, Fig. 6.64, have matching permanent magnets in the inner ring, thus
driven speed equals driving speed whenever the torque is within the coupling's capability.

The choice of magnetic drive type has been influenced by advances in
magnetic technology. Hatch [6.31], apparently speaking for earlier technology, argued that the speed advantage of synchronous drives is too small to
compensate for the lower weight, lower cost, and higher starting torque of
eddy current drives. More recently, Minter [6.32] observed that eddy current
drives are only effective at speeds of 3000 r/min and higher, that continuous
operation at high torque (hence high slip) can lead to overheating and failure,
and that the necessarily very thin containment can preclude high pressure
applications. For these reasons, Minter noted, synchronous drive has superseded eddy current for all but high temperature (above 400°F) and noncriti-
cal, low cost applications. Nasr [6.33] dealt exclusively with synchronous
drives.

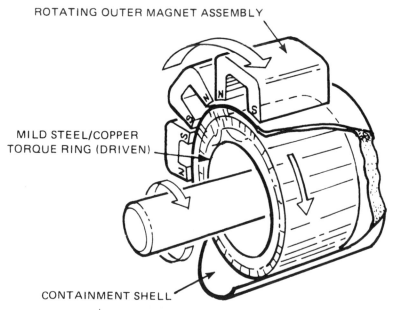

ROTATING OUTER MAGNET ASSEMBLY

MILD STEEL/COPPER
TORQUE RING (DRIVEN)

CONTAINMENT SHELL

FIG. 6.63.   Eddy current magnetic drive. (Courtesy The Kontro Co., Inc.)

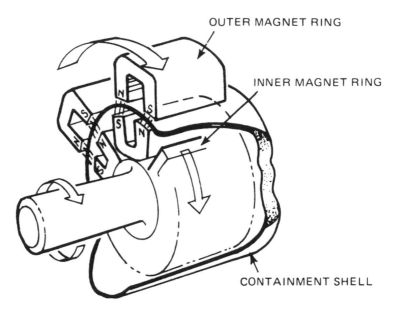

OUTER MAGNET RING

INNER MAGNET RING

CONTAINMENT SHELL

FIG. 6.64. Synchronous magnetic drive. (Courtesy The Kontro Co., Inc.)

Both drive types incur losses beyond those in conventional pumps: disc friction from rotation of the inner ring in the pumped liquid, magnetic drag between the outer ring and the containment can when it's metallic and, in eddy current drives, the power loss produced by slip across a constant torque device.

Canned Motors.   Close coupled electric motors within the pressure boundary come in a number of configurations. For chemical process service, where corrosion is generally a factor, the usual configuration is a dry pit version of the canned wet rotor, dry stator rotor, see Fig. 6.65. In this configuration the pressure boundary is between the motor rotor and stator, allowing the stator to run dry and avoiding the need for pressure sealing of the stator leads. A notable departure from usual practice is the use of wet pit, wet rotor, wet stator motors in noncorrosive cryogenic service.

Canned motors are less efficient than comparable conventional motors. Additional losses incurred are higher "windage" due to the rotor running in liquid (offset to some extent by rotor proportions) and the greater effective air gap needed for the can.

Rotor Bearings.   The second basic problem in hermetically sealed pumps is supporting and locating the pump rotor. By employing the correct form of hydraulic design and because no part of the motor is exposed to atmosphere, rotor loads, radial and axial, can be kept relatively low. In current designs these low loads are carried by journal and thrust bearings of wear-resistant materials, lubricated by the pumped liquid. Under favorable circumstances

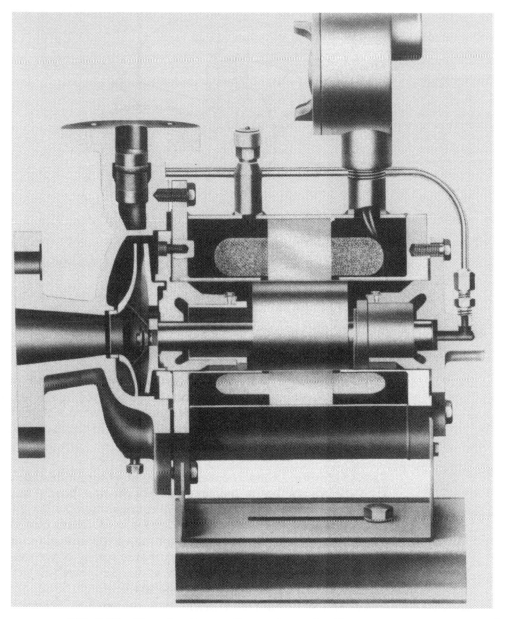

FIG. 6.65. Canned motor pump for corrosive service. (Courtesy Crane Co.)

this solution yields acceptable life, but when the pumped liquid contains solids or lacks lubricity, the life of standard bearings can be short. Special designs are being developed though; Minter reports magnetic drive applications on abrasive slurries containing up to 60% solids.

Limitations.   Typical application limits for hermetically sealed pumps are summarized in Table 6.2.

TABLE 6.2 Typical Application Limits for Hermetically Sealed Pumps

| | Pump Type | |
Condition	Magnetic Drive	Canned Motor
Pressure (usual), lb/in.²gauge	300	300
Pressure (special), lb/in.²gauge	7250	5000
Temperature (uncooled), °F	750	400
Temperature (cooled), °F	850	1000
Power, hp	700	2000
Explosion proof	Easy	Difficult

Magnetic drive pressure/temperature limits are for eddy current drives and metallic containment cans. Hatch notes the availability of small nonmetallic pumps good for 150 lb/in.²gauge at 300°F and rated for approximately 5 hp.

The temperature limit for cooled, canned motor pumps is based on pumps with a thermal barrier to minimize heat transmission to the motor and some form of stator cooling.

Special Problems.   Starting torque with synchronous magnetic drives may not be sufficient for pumps handling thixotropic liquids.

Containment can material and construction are critical. The material must have high electrical resistivity (to lower drag in magnetic drives) and high corrosion resistance. Compounding these, the can has to be thin (0.06 in. for synchronous magnetic drives, thinner for eddy current drives and canned motors) so its manufacture is difficult. Minter reports a marked improvement in can reliability upon changing from welded to formed construction.

Solids-laden liquids need care. As already noted, bearing life is one problem. The other is damage to the containment can should large solids get into the space between it and the rotor.

Vapor or gas accumulation at high points in horizontal or vertical motor up orientations has caused two problems. The first is bearing failure due to running dry. The second is differential cooling distorting the containment can, leading to rubbing contact with the rotor and consequent can failure.

Bearing wear will eventually allow rubbing contact within the pump. In small, synchronous drive pumps this will result in rotor seizure but little else. Eddy current drive pumps, however, will quickly overheat and may suffer catastrophic failure. Canned motor pumps have a close clearance between the rotor and can. Compounding this, the can is very thin and prone to catastrophic rupture should the rotor rub against it. Rayner and Waughman [6.34] cite nonpenetrating vibration probes as a means of avoiding this problem.

6.20.3. Induced Vortex Pumps

Instead of imparting energy to the pumped liquid by direct action of their impeller vanes, as in conventional centrifugal pumps, induced vortex pumps

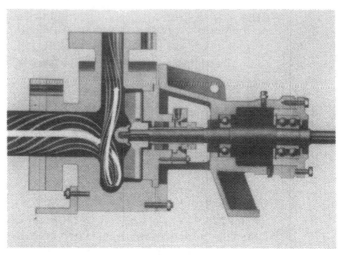

FIG. 6.66. Induced vortex or "torque flow" pump. (Courtesy Met-Pro Corp./FYRBROC Division.)

impart energy by momentum exchange. Figure 6.66 shows a typical liquid end. The impeller is recessed back out of the pumped flow to give a free flow path. Low momentum liquid enters the impeller near the hub, is accelerated through the impeller, then issues as high momentum liquid into the pumped flow where it gives up energy to the flow.

Induced vortex pumps (also known as "recessed impeller" or "free flow" pumps) are better able than conventional centrifugal pumps to handle pumpage containing large solids, stringy material, or gas. Given the nature of energy transfer, the efficiency of induced vortex pumps is lower than that of equivalent, conventional centrifugal pumps. Their performance characteristics, however, are quite similar.

Induced vortex liquid ends are used in horizontal and vertical pumps. Shaft seals and bearings are the same as those for conventional end suction chemical process pumps.

6.20.4. Regenerative Turbine Pumps

Before the advent of high speed, single stage pumps, another alternative for low flow, high head duties was the regenerative turbine pump. Figure 6.67 shows a typical liquid end. The essential feature of this design is a multiple vaned impeller running in a casing shaped to cause the liquid to continually re-enter the impeller during its passage around the casing. With multiple passes through the impeller, the pump is able to develop more head than a conventional centrifugal pump of the same capacity and speed. A close clearance "stripper block" in the casing stops the re-entry process and separates suction and discharge. To eliminate axial thrust on the rotor, the impeller and casing are double sided, and the impeller is free to move axially, thus allowing it to find a "balanced" position within the casing.

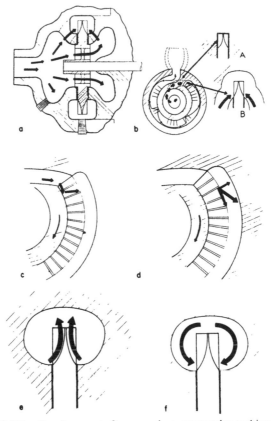

FIG. 6.67. Development of pressure in a regenerative turbine pump.

Figure 6.68 shows typical regenerative turbine pump performance characteristics. Both head and power rise steeply with decreasing flow, reaching a maximum at shutoff. To avoid motor overloading, regenerative turbine pumps require some form of minimum flow bypass. A conventional pressure relief valve is the usual arrangement.

Regenerative turbine pumps require a close clearance between the impeller and "stripper block," typically 0.002 in. per side, for effective operation. The need for such a close clearance makes the pump type very sensitive to wear, and thus suitable only for clean services.

Providing the casing is filled initially and the air or gas can be discharged, regenerative turbine pumps are self-priming. Their tolerance of two-phase flow means a gradual performance degradation with decreasing NPSHA. To account for this, each size pump will have several rating curves, each for a different value of NPSHA.

Shaft seals, rotors, and bearings are similar to those for conventional chemical process pumps. Rotor stiffness, however, has to be higher to accommodate the high radial thrust inherent in the design.

Given their performance characteristics, relatively low efficiency, intolerance of solids, and the availability of alternatives, regenerative turbine pumps are no longer widely used.

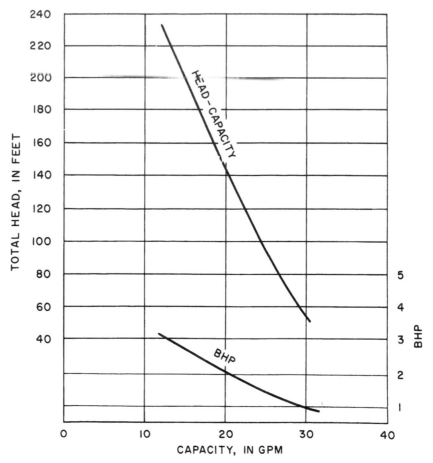

FIG. 6.68. Typical performance characteristics of a regenerative turbine pump.

6.21. Materials and General Design

The broad considerations for materials are corrosion resistance, strength, stiffness (not the same as strength), and toughness, all at the service temperature. In addition, the material must retain adequate toughness at ambient temperature lest it prove difficult to fabricate and install.

To a great degree, materials and general design are interrelated, so there is justification in discussing the two together.

One of the virtues of centrifugal pumps is their availability in a very wide range of materials. The materials generally available for the types of centrifugal pumps used in chemical processing are shown in Table 6.3.

Many of the materials listed in Table 6.3 warrant additional comment.

TABLE 6.3 Centrifugal Pumps Available (A) for Chemical Processing Materials

Material (common name)	Chemical (ANSI)	Process (API)	Slurry	Elbow	Between Bearings, Multistage, High Speed
Metals					
Carbon steel		A			
Ductile iron	A				
316 stainless steel	A	A		A	A
Alloy 20	A	A		A	A
CD4MCu	A			A	A
Hastealloy	A			A	
Duplex alloys	A			A	
Titanium	A				A
Zirconium	A				
Silicon iron	A				
Chrome iron			A		
Nonmetal					
Rubber			A		
Plastic, thermoplastic	A				
Plastic, thermoset	A				
Glass	A				
Ceramic	A				
Graphite	A				

Alloy 20 and higher alloys tend to have very specific corrosion performance, i.e., a small change in liquid constituent can markedly alter serviceability. Further, these metals require particular care in manufacture if they're to perform as expected.

High silicon iron offers excellent corrosion and abrasion resistance, but poses special problems in casting (propensity to crack) and machining (high hardness). The advent of CBN (cubic boron nitride) tools and laser arc machining tends to alleviate the latter problem.

Chrome iron, up to 35%, offers a good balance of corrosion and abrasion resistance. Its popularity suggests it is easier to cast than high silicon iron. Pumps made in either high silicon or chrome iron are of a special design to minimize machining.

Plastic fall into two categories: thermoplastic and thermoset. Thermoplastics have broad chemical resistance but generally lack strength and so are only used as linings in pump construction. Thermoset plastics, with the exception of fluorocarbons, have narrower chemical resistance but at the same time sufficient strength for use as whole components, see Fig. 6.69. Fluorocarbons, while generally thermoset plastics, perform more like thermoplastics; they have very broad chemical resistance and unless strengthened with fillers, can only be used as linings.

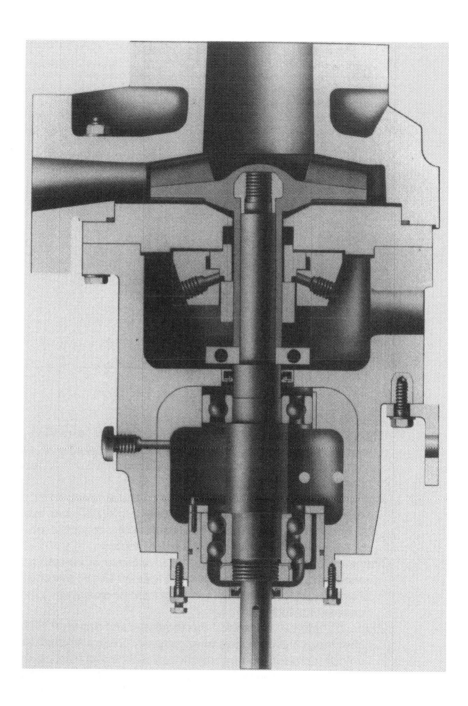

FIG. 6.69. Nonmetallic pump for corrosive service. (Courtesy Worthington Pump, Dresser Industries, Inc.)

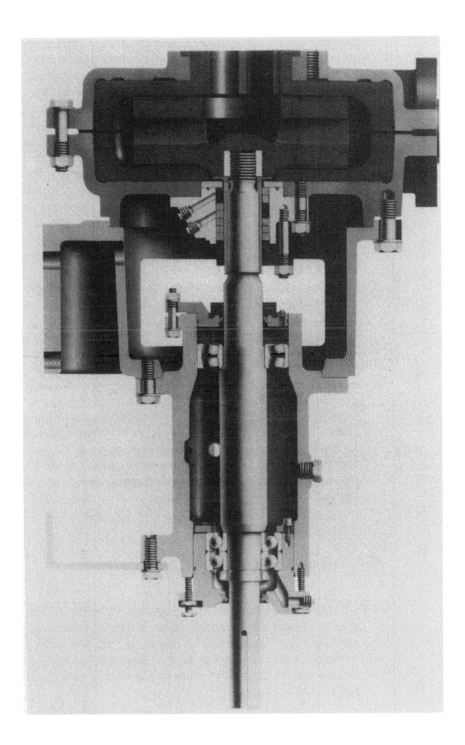

FIG. 6.70. Ceramic-lined slurry pump for high temperature erosive service. (Courtesy Worthington Pump, Dresser Industries, Inc.)

Rubber offers excellent corrosion and abrasion resistance. Its use in chemical processing is limited by service temperature (250°F (120°C) maximum for synthetic) and strength. Like thermoplastics, rubber is used as a lining.

Glass is essentially inert. While immensely strong, it lacks toughness so it can only be used in pumps as a lining. Birk and Peacock noted in Ref. 1.1 that glass linings are prone to fine cracking, a consequence of differential thermal expansion, which allows corrosion of the metal beneath it.

Ceramic, a broad term, is finding use in corrosive/erosive services at temperatures beyond the limit for rubber. A common ceramic for this application is silicon nitride bonded silicon carbide. At present, ceramics have low strength and toughness, which limits their use for whole parts to the impeller. Casings are made using a constrained ceramic liner, see Fig. 6.70. In sealless pumps, ceramics such as aluminum oxide and silicon carbide are used for pump bearings and, occasionally, the entire rotor.

Graphite, suitably treated to render it impervious, is employed in the most corrosive services. Being soft, it has no useful erosion resistance. The construction of graphite pumps is similar to that for ceramic, see Fig. 6.70, and for the same reasons.

Coatings and surface treatments offer equal or improved performance plus conservation of valuable materials. At this stage, abrasion-resistant coatings for seals and running clearances are reasonably well established. Surface conversion treatments for abrasion resistance have not been widely used in pumps. Those involving diffusion are generally incompatible with corrosion-resistant alloys (sensitization or embrittlement), but others, such as ion implantation, hold promise. All the notable surface coatings or treatments are relatively thin, thus they cannot be considered sacrificial. Improving corrosion resistance by coating has met with mixed results; microporosity allows corrosion of the substrate and often results in spalling of the coating. Progress is being made though. Yeaple [6.35], for example, reports the development of a coated antifriction bearing said to have the corrosion resistance of "stainless steel," presumably Type 316, without sacrificing bearing capacity.

6.22. General Design

Reliable pressure containment is of prime importance. The casing design should include a corrosion/erosion allowance consistent with its envisaged service. Manufacture of the casing should be to an appropriate national standard and to a quality level consistent with its criticality. Some designs may be easier to produce to consistent quality levels as weldments rather than as castings.

To save exotic material, many designs use a separate clamping ring, or the bearing adaptor flange, to retain the casing cover, see Fig. 6.71. The clamping ring is part of the pressure containment and should be made of a ductile material lest it fracture during thermal or mechanical shock.

Gaskets must be compatible with the pumped liquid and should be mechanically confined against blow-out. Figure 6.71 shows the configuration.

Wherever possible, locating fits should be dry to avoid minor corrosion destroying concentricity, see Fig. 6.71.

Avoid wetted, stagnant regions, particularly in parts under tension. Not doing so risks premature failure by concentration cell corrosion or stress cracking or both. Problem areas are impeller attachment, locating fits, wearing rings or wear plates, and threaded connections. Figures 6.71 and 6.72 illustrate the regions.

Impeller attachment sealing, Fig. 6.72(a), should use either hard gaskets or O rings to ensure the assembly remains tight during operation.

Avoid regions where centrifuge action could raise the local solids content, leading to erosion of the outer surface. Figure 6.71 shows such a region and offers guidance on its solution.

Pumps for erosive service should have as few joints as possible and no tapped connections into the casing; each joint and connection is an erosion site.

In flexible rotor multistage pumps, materials for internal bushings must be gall resistant and of reasonable bearing capacity. Teflon has proven serviceable in small pumps. Ceramic coatings and filled graphite have given good results in larger pumps.

Thermoset plastics, while strong enough for whole pump parts, are an order of magnitude less stiff than metals and, unless filled, have low toughness. Parts designed for plastic therefore have to be proportioned and fabricated by taking account of the material properties. Although this seems obvious, it is frequently overlooked.

Some services require that the bearing housings and the connection to the casing be corrosion resistant. Type 316 stainless steel can be used in place of iron. Plastic is an alternative, but the design would need to take account of stiffness and toughness (see above), heat dissipation, and would require metal liners to locate antifriction bearings.

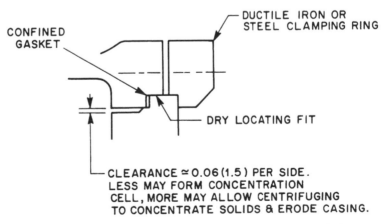

FIG. 6.71. Desirable features in radially split casing closure design.

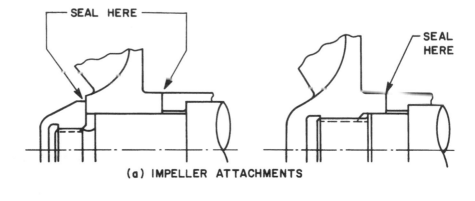

(a) IMPELLER ATTACHMENTS

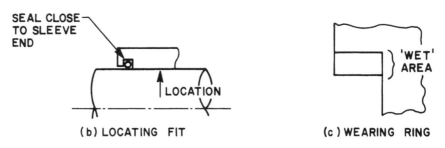

(b) LOCATING FIT   (c) WEARING RING

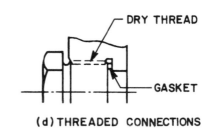

(d) THREADED CONNECTIONS

FIG. 6.72.   Construction features to reduce risk of local corrosion in chemical pumps.

6.23. Material Selection

Selecting materials is a major task. The first rule is to try to learn what has been used before for the same service, or failing that, a similar service. The second rule is to search out any new developments in materials that might be applicable to the service.

Table 6.4 lends weight to the second rule. Presented by Byrd [6.20], this table shows a departure from traditional metals for severe services and at the same time recognizes the current limitations of new materials. Being drawn from plant experience, the table tends to be conservative. Plastic pump development, for example, is evolving. Banerjee et al. [6.36] report a fiber-

TABLE 6.4 Centrifugal Pumps for Chemical Processing Material Selection Guide—General[a]

	Material					
Duty	STL D. Iron	316 and Special Alloys	Plastic	Common Rubbers	Glass or Ceramic	PTFE, Graphite, Viton
Mildly corrosive liquids	×	×	×	×	×	×
Hydrocarbons (e.g.,benzene and alcohol)	×	×	?	?	×	×
Chlorinated hydrocarbons with trace HCl	×	×	*	*	×	×
Alkali (e.g., NaOH)	×	×	×	?	*	×
Acids (e.g., HCl)	*	?	×	?	×	×
Physical stress (e.g., thermal shock, piping loads)	×	×	*	×	*	?

[a]× = acceptable, ? = borderline, * = not acceptable.

reinforced vinyl ester pump with broad chemical resistance to acid and alkaline solutions, and with piping load capability equal to ANSI metal pumps.

References

6.1. W. H. Fraser and E. P. Sabini, "The Effect of Specific Speed on the Efficiency of Single Stage Centrifugal Pumps," in *Proceedings of the 3rd International Pump Symposium*, Texas A&M, 1986, pp. 55–59.

6.2. D. J. Vlaming, "Analysis of Cavitation Provides Advanced NPSH Estimates for Centrifugal Pumps," *Oil Gas J.*, October 19, 1984.

6.3. B. Schiavello, "Visual Study of Cavitation—An Engineering Tool to Improve Pump Reliability," *EPRI 1st International Conference on Improved Coal-Fired Power Plants*, Palo Alto, California, November 19–21, 1986.

6.4. G. F. Wislicenus et al., "Cavitation Characteristics of Centrifugal Pumps Described by Similarity Considerations," *ASME Trans.*, pp. 17–24 (January 1939).

6.5. G. F. Wislicenus, *Fluid Mechanics of Turbomachinery*, 2nd ed., Dover Publications, New York, 1965.

6.6. W. H. Fraser, *Recirculation in Centrifugal Pumps*, Presented at the Winter Annual Meeting of ASME, Washington, D.C., November 15–20, 1981.

6.7. V. S. Lobanoff and R. R. Ross, *Centrifugal Pumps—Design and Application*, Gulf Publishing, Houston, Texas, 1985.

6.8.   J. H. Doolin, "Judge Relative Cavitation Peril with Aid of These 8 Factors," *Power Mag.*, pp. 77–80 (October 1986).

6.9.   R. R. Ross and P. H. Fabeck, "Flexible Inducer Pumps Can Have Wide Operating Range," *Oil Gas J.*, August 5, 1985.

6.10. J. H. Doolin, "Select Pumps to Cut Energy Costs," *Chem. Eng.*, January 17, 1977.

6.11. D. Konno and Y. Yamada, "Does Impeller Diameter Affect NPSHR?," in *Proceedings of the International Pump Symposium*, Houston, Texas, 1984, pp. 29–35.

6.12. S. Yedidiah, "Factor Size and Speed into NPSHR Comparisons," *Power*, p. 78 (June 1973).

6.12a. A. J. Stepanoff, *Pumps, Blowers, Two Phase Flow*, Wiley, 1965.

6.12b. S. Yedidiah, "Factor Size and Speed into NPSHR Comparisons," *Power*, p. 78 (June 1973).

6.13. I. J. Karassik, L. L. Petraccaro, and J. T. McGuire, *Variable Frequency Pump Drives*, Presented at the 8th Annual International Energy Technology Conference & Exhibition, Houston, Texas, June 17–19, 1986.

6.14. A. Agostinelli et al., "An Investigation of Radial Thrust in Centrifugal Pumps," *Trans. ASME Int. Basic Eng.*, Paper #59-Hyd-2.

6.15. A. A. Gasiunas, "Development of a Single-Stage Boiler Feed Pump for Nuclear Power Stations," in *Proceedings of the Institute of Mechanical Engineers Conference*, 1970, Paper 184-3N.

6.16. P. Hergt and P. Krieger, "Radial Forces in Centrifugal Pumps with Guide Vanes," *Proc. Inst. Mech. Eng.*, *184*, Pt 3N, 101–107 (1969–70).

6.17. N. Uchida et al., "Radial Force on the Impeller of a Centrifugal Pump," *Bull. JSME*, *14* (76), 1106–1117 (1971).

6.18. H. Kanki et al., *Experimental Research on the Hydraulic Excitation Force on the Pump Shaft*, ASME, Paper 81-OE7-71.

6.19. M. Grohmann, "Extend Pump Application with Inducers," *Hydrocarbon Processing, 9* (12), 121–124 (1979).

6.20. G. C. M. Byrd, *Pump Users Experience, Pumps and Pumping: A Practical Guide to Recent Developments*, Institute of Chemical Engineers, Manchester, UK, November 1985.

6.21. ANSI B73.1M-1987, *Specification for Horizontal End Suction Centrifugal Pumps for Chemical Process*, American National Standards Institute, New York.

6.22. ISO-2858.

6.23. DIN-24256.

6.24. API-610, 7th ed., American Petroleum Institute, Washington, D.C., 1989.

6.25. H. P. Bloch, "Mechanical Reliability Review of Centrifugal Pumps for Petrochemical Service," in *Proceedings of the ASME Failure Prevention and Reliability Conference*, Hartford, Connecticut, 1981.

6.26. H. H. Anderson, "Reliability Assurance in Pumps," in *Proceedings of the BPM 9th Technical Conference on Reliability—The User–Maker Partnership*, April 1985, pp. 93–101.

6.27. ANSI B73.2M-1987, *Specification for Vertical In-Line Centrifugal Pumps for Chemical Process*, American National Standards Institute, New York.

6.28. BS4082-1966.

6.29. A. B. Duncan and J. F. Hood, "The Application of Recent Pump Developments to the Needs of the Offshore Oil Industry," in *Proceedings of the Conference on Pumps and Compressors for Offshore Oil and Gas*, London, UK, June 29–July 1, 1976, pp. 7–24.

6.30. V. S. Lobanoff and R. R. Ross, *Centrifugal Pumps, Design and Application*, Gulf Publishing, Houston, Texas, 1985.

6.31. J. A. Hatch, *Pump Handbook*, 2nd ed. (Karassik, I.V. Krutzsch, W. C. Fraser, W. F. and Messina, J. P. eds.), McGraw-Hill, New York, 1986.

6.32. P. Minter, "Glandless Pumping," in *Pumps and Pumping, A Practical Guide to Recent Developments*, Institute of Chemical Engineers, Manchester, UK, November 1985.

6.33. Nasr, A. M., "When to Select a Sealless Pump," Chemical Engineering, 93 (10), 85–89 (May 26, 1986).

6.34. K. G. Rayner and L. G. Waughman, "A 150 kW Glandless Canned Motor Pump for Biochemical Plant Service," in *Proceedings of the BPMA 9th International Technical Conference on Reliability, The User–Maker Partnership*, Coventry, UK, April 1985, pp. 189–202.

6.35. F. D. Yeaple, "Material Choices Are Critical for Best Mechanical Performance and Life," *Prod. Eng.*, *49*(3), 57–88 (March 1978).

6.36. B. R. Banerjee et al., "Development of a Polymer-Composite Industrial Machine," in *Pumps and Pumping, A Practical Guide to Recent Developments*, Institute of Chemical Engineers, Manchester, UK, November 1985.

7. Rotary Pumps

If a single service is desired to define rotary pump usage, it is to pump viscous liquids. The thrust of their development has been to devise mechanisms able to pump viscous liquids with good efficiency. Side developments have added embelishments to afford some tolerance of lack of lubricity and abrasive solids in the pumped liquid.

7.1. Fundamental Operation

A great many rotary pump mechanisms have been developed. All, however, function in the same basic manner.

A series of fixed or controlled displacement volumes move continually and at essentially constant speed within the pump.

Movement of the displacement volumes is either rotary or translational, but always produced by rotary motion.

As a volume passes through the suction region, it fills with liquid.

Further movement captures the volume within close clearances and brings it to the discharge region.

In the discharge region the captured volume is gradually "displaced" by meshing, eccentricity, or plunger action.

Figure 7.1 illustrates the basic function, in this case for an external gear pump, which rotates the "captured" volume and displaces it by "meshing."

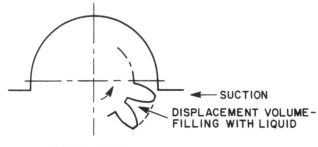

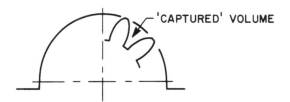

(a) SUCTION

(b) MOVING 'CAPTURED' VOLUME
(ROTATING)

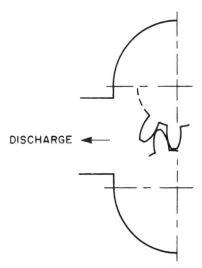

DISCHARGE ◄—

(c) DISPLACING 'CAPTURED' VOLUME
(MESHING ACTION)

FIG. 7.1.   Basic rotary pump action (external gear pump).

The essential feature of rotary pumps is movement of the displacement volumes from suction to discharge, thus eliminating the need for suction and discharge valves.

With the wide variety of rotary pump mechanisms, it is difficult to speak of basic parts, so that will be left until the introduction of the various types.

7.2. General Characteristic

Rotary pumps are positive displacement devices. Thus, for a given speed and viscosity, they produce a nearly constant flow, the variation depending upon the differential pressure across the pumps; see chapter 5. With such a flow characteristic, the practice is to show rotary pump performance characteristics with differential pressure as the independent variable. A typical performance characteristic is shown in Fig. 7.2.

The departure of actual flow from theoretical with increasing ΔP is known as "slip." In that the difference represents the flow which has "slipped by" the close internal clearances, the term is correct. Slip in rotary pumps in analogous to internal leakage in centrifugal pumps. The ratio of flow delivered/flow displaced internally is the "volumetric efficiency." How slip varies with differential pressure depends upon the particular design.

Power absorbed starts with a fixed loss, the friction involved in moving the internal parts, then increases with differential pressure.

Flow and power versus pressure are the working characteristics of rotary pumps. The ratio of their output, as a power, to the power absorbed gives "pump efficiency" or, as it's often called, "mechanical efficiency." Pump efficiency is the product of volumetric and mechanical efficiencies, the latter related to internal friction losses, so it's better not to confuse the terms. In determining the worth of a particular design, pump efficiency is the important parameter.

7.3. NPSHR

Consistent with the principle noted in chapter 4, rotary pumps require a certain amount of net positive suction energy (NPSH) to produce their rated

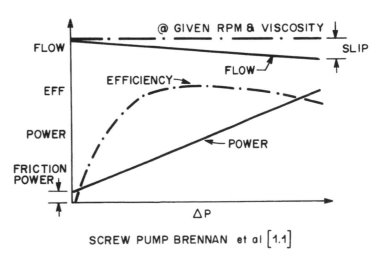

SCREW PUMP BRENNAN et al [1.1]

FIG. 7.2. Typical rotary pump performance characteristic.

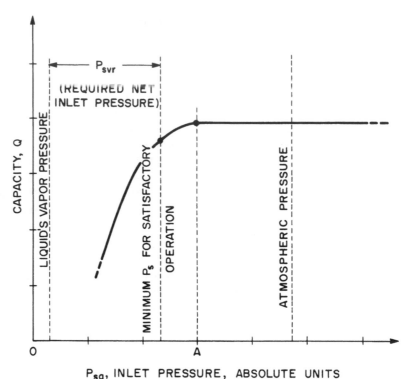

FIG. 7.3.   Variation of flow with suction pressure; other operating conditions constant.

capacity. If NPSHA is less than NPSHR, vaporization in the suction region will prevent complete filling of the moving "displacement volumes," causing a drop in capacity, see Fig. 7.3. Because the capacity is practically constant for a given speed and viscosity, the NPSHR for the general characteristic in Fig. 7.2 is a single value. By current convention, NPSHR for rotary pumps is generally expressed in pressure units.

7.4. Effect of Viscosity

As viscosity increases, internal leakage (slip) decreases, thus raising the capacity delivered. At the same time, friction power increases, thus raising the absorbed power. Figure 7.4 shows these two effects qualitatively.

Determined over a wide range of viscosities, efficiency shows a peak, with lower viscosities producing a rapid fall and higher viscosities a gradual decline, see Fig. 7.5. Little, in Ref. 1.1, attributes the rapid fall with lower viscosity to the volumetric efficiency falling below 50% as slip increases.

For a given flow, NPSHR increases with viscosity. When NPSHA is limited, the higher NPSHR associated with increasing viscosity limits flow, hence pump speed. Figure 7.6 shows data presented by Dolby and Carter [7.1] for helical rotor pumps.

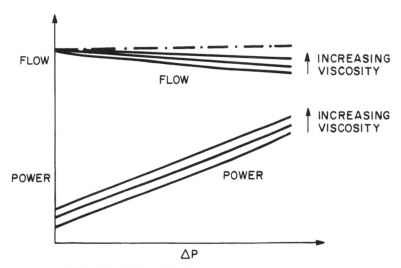

FIG. 7.4. Effect of liquid viscosity on flow and power.

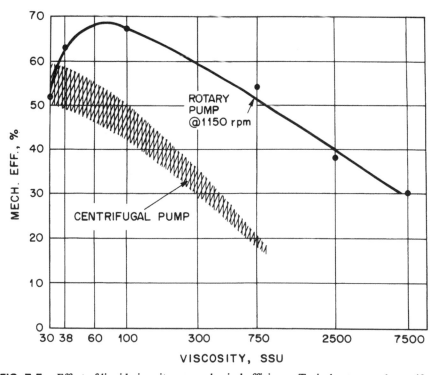

FIG. 7.5. Effect of liquid viscosity on mechanical efficiency. Typical rotary and centrifugal pumps rated 40 gal/min at 50 lb/in.²

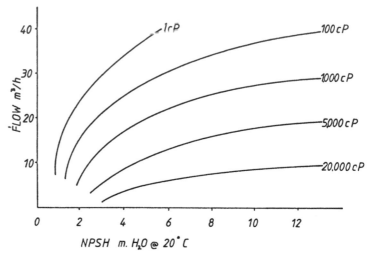

FIG. 7.6. Increase in NPSHR with increasing liquid viscosity. Typical helical rotor pump.

7.5. Effects of Gas

Gas in the pumped liquid is compressed at the pump discharge, thus reducing the delivered capacity. Small amounts of dissolved gas produce only a reduction in capacity, the reduction depending on both the gas content and the pump's suction pressure, see Fig. 7.7, and so are tolerable. Larger volumes of entrained gas (beyond 15% by volume per Davidson [1.5]) can cause noise and vibration, and therefore limit the types of rotary pump that can be used. With appropriate design (progressive compression and no recirculation from discharge to suction), gas volumes up to 95% have been handled.

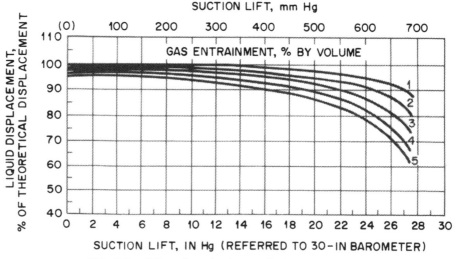

FIG. 7.7. Effect of entrained gas on liquid displacement.

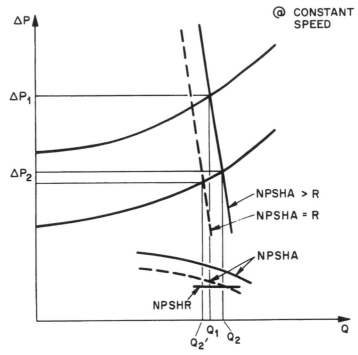

FIG. 7.8.   Rotary pump interaction with system.

7.6. Interaction with System

The principle governing rotary pump interaction with the system is the same as that for centrifugal pumps: the operating capacity is given by the intersection of the pump and system head (or pressure) characteristics. Figure 7.8 shows such an interaction. What is distinctly different in practice, however, is that the very steep pressure characteristic of the rotary pump means there is very little change in capacity as the system resistance varies.

Again, as is the case for centrifugal pumps, the actual operating capacity can be influenced by the NPSHA. If the NPSHA is equal to or below that required, the actual pump capacity will be lower than given by the rating curve. Figure 7.8 also illustrates this.

7.7. Parallel and Series Operation

Parallel operation is quite straightforward; capacities are added at equal pressure differentials to produce a combined characteristic, and the pumps operate at the intersection of that and the system characteristic. The only caution is that the pumps be sufficiently similar to flow share. Figure 7.9

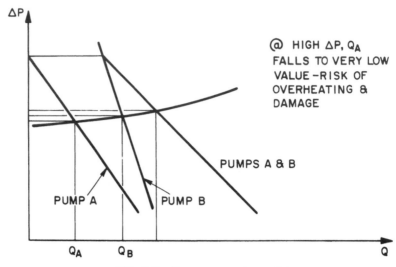

FIG. 7.9. Rotary pumps in parallel.

shows an extreme case where one pump has a "softer" pressure characteristic (lower flow regulation) than the other, thus posing the risk of low flow through the "softer" pump if the pair is run at too high a pressure differential.

Given the basic character of displacement pumps, their operation in series is not a general practice. Circumstances which may warrant it are driver or pump power limitations and, in those pumps capable of handling abrasive solids, a need to limit differential pressure for longer wear life. The difficulties with series pumping are two. First, the pumps pass the same flow, and unless their pressure characteristics are fairly soft or their control very precise, small differences in pump geometry or speed can mean wide variations in differential pressure, see Fig. 7.10. Second, succeeding pumps have to withstand higher suction and working pressures.

7.8. Varying Flow

Because their flow regulation is very high, sensible flow variation in rotary pumps can only be achieved by bypassing or varying speed.

Bypassing, see Fig. 7.11, involves returning part of the delivered capacity to the pump's suction vessel (continually returning flow to the pump's suction should be avoided since it can cause overheating). The bypass valve has to be sized for the maximum flow likely to be bypassed, and its pressure drop is essentially the differential across the pump. As process flow varies, pump gross capacity is practically constant and pump power changes only by the variation in system resistance. Bypassing, therefore, is not efficient. For small pumps or small bypass capacities, simplicity and ease of operation offset the energy loss. For larger pumps the energy loss may warrant variable speed operation.

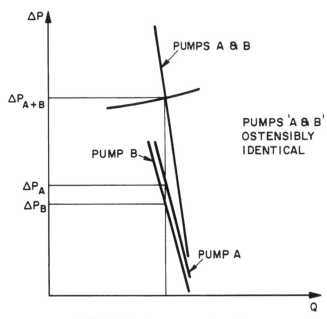

FIG. 7.10. Rotary pumps in series.

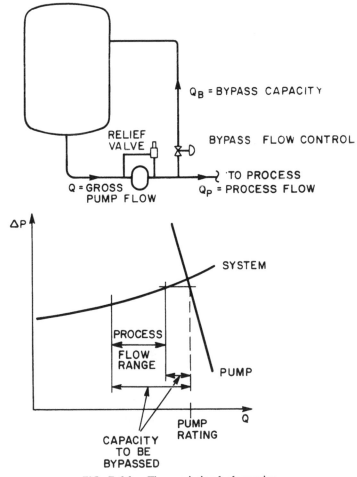

FIG. 7.11. Flow variation by bypassing.

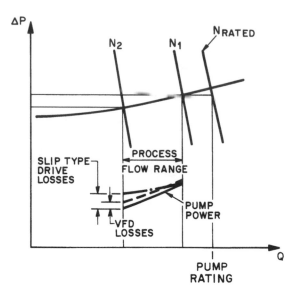

FIG. 7.12.   Flow variation by varying speed.

Varying the pump speed is usually realized by interposing a variable speed drive between the pump and its motor or varying the power supply frequency to the electric motor. Variable frequency drive (VFD), as it's known, offers greater energy savings and is gaining acceptance now that its reliability has been demonstrated.

Variable speed operation is shown in Fig. 7.12; pump speed is simply adjusted to produce the required process flow. Pump power decreases with flow and system pressure. Drive evaluation, however, has to take account of drive losses. When the process flow range is high, hence the speed ratio is high, the higher losses in a slip-type drive tend to offset the higher capital cost of VFD.

Other means of variable speed drive such as steam turbine, hydraulic, or pneumatic motor or engine are, of course, quite feasible. They are, however, rarely used in comparison to the simple squirrel cage electric motor.

7.9. Pump Selection

Rotary pump development has created myriad variations of the basic concept. The selection of a particular variation depends upon the service conditions. The particular variation, in turn, can have an effect on the system and pump rating. Figure 7.13 shows the basic process for making and checking a selection.

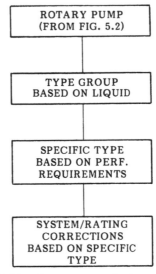

FIG. 7.13. Rotary pump selection process.

7.10. Nature of Liquid

All the variations of a rotary pump can be assigned to one of three groups, based on the nature of the pumped liquid. Four characteristics of the liquid serve to define the type groups:

Rheological characteristic
Viscosity and lubricity
Presence and nature of solids
Presence of entrained gas

The three rotary pump groups and the broad nature of the liquids for which they are suitable are listed in Table 7.1.

TABLE 7.1 Rotary Pump Groups and the Liquids for Which They Are Used

Group	Nature of Liquid
1. Internal mechanism support and force transmission.	Medium viscosity, Newtonian, good lubricity, free of solids and entrained gas
2. External mechanism support and force transmission.	Medium viscosity, Newtonian, poor lubricity, nonabrasive solids, entrained gas (in some designs)
3. Elastomer element in mechanism.	Medium to high viscosity, non-Newtonian, abrasive solids

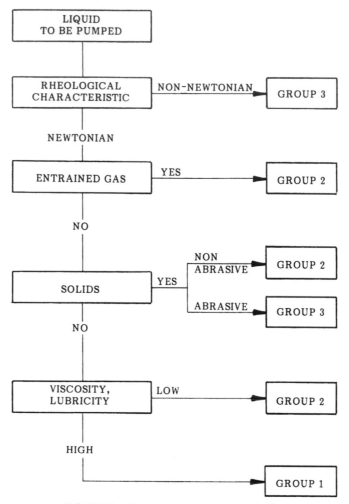

FIG. 7.14.   Rotary pump type group selection.

Figure 7.14 illustrates how the most appropriate rotary pump group can be chosen from the liquid characteristics.

7.11. System Corrections

7.11.1. Rated Flow

With displacement pumps the usual practice is to rate the pump for the required process flow, i.e., not to add any margin. The virtue of zero flow margin is that it avoids having to provide means to vary the flow. When process flow can vary or the service is likely to cause rapid pump wear, a flow margin is justified. The amount depends upon the likely flow variations in the first case and the desired time between overhauls in the second.

TABLE 7.2   Typical Flow Pulsations

Pump Type (see description)	Flow Pulsation, % of Mean
Any with inlet dampener	0
Screw (2 and 3)	0
Gear (helical)	0
Vane	5
Lobe: 2 lobe	25
3 lobe	15
Helical rotor	3
Peristaltic: 2 roller, planar cam	100
3 roller	50

7.11.2. Flow Pulsation

While it has long been recognized that some designs of rotary pumps produce pulsating flow, most system design practice makes no allowance for such a flow. This suggests that in general, rotary pump flow pulsations do not manifest themselves as serious problems with NPSHA or pressure pulsations. The form of the flow pulsations and the characteristics of the liquid may well account for this.

For cases where flow pulsations are considered likely to cause trouble or are suspected as the cause of trouble, Davidson [1.5] provides data for estimating consequent pressure pulsations and acceleration head. As a guide to potential problems, typical flow fluctuations for the usual pump types from Davidson [1.5] are shown in Table 7.2.

Installing an inlet dampener effectively eliminates any problem. Dampener selection should be made by the manufacturer, with full knowledge of the form of pump flow pulsation, and thus must follow selection of the pump type.

7.12. Pump Type

Figure 7.15 shows the types of rotary pump commonly used for chemical processing. In Fig. 7.15 the pump types are categorized as sealed (having a shaft through the pressure boundary) and sealless (no shaft through the pressure boundary.) Each of the categories is divided into the type groups discussed above. Group 2 pumps cannot, by definition, be sealless.

Typical performance capabilities of the types in Fig. 7.15 are shown in Table 7.3. These data are, of necessity, only nominal. Individual designs may exceed the nominal, and design evolution can move the nominal, so the table can serve only as a guide.

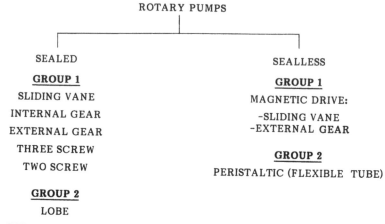

FIG. 7.15. Rotary pump types for chemical processing.

TABLE 7.3 Rotary Pumps for Chemical Processing: Typical Performance Capabilities

Group	Type	Flow (gal/min)	Differential Pressure (lb/in.²)	Viscosity (SSU)	Temperature (°F)
Sealed:					
1	Sliding vane	1,000	125	500,000	450
	Internal gear	1,000	250	1,000,000	650
	External gear	2,000	300	1,000,000	650
	Triple screw	3,000	3,000	50,000	200
	Twin screw	4,000	2,000	1,000,000	300
2	Lobe	400	450	1,000,000	500
	External gear	2,000	300	1,000,000	650
	Twin screw	10,000	2,500	1,000,000	850
3	Flexible vane	100	30	100,000	180
	Helical rotor	1,250	300	1,000,000	200
Sealless:					
1	Magnetic drive:				
	Sliding vane	60	15	N.D.	N.D.
	External gear	60	100	35,000	N.D.
2	Peristaltic	300	220		180

Internal (vane-in-rotor) pump.

FIG. 7.16. Sliding vane pump. (Reprinted with permission from Ref. 1.1.)

7.13. Sliding Vane

The "displacement volumes" are formed between vanes in the rotor, the rotor o.d., and the casing bore, see Fig. 7.16. Displacement of the volumes is produced by a profiled casing bore. The vanes slide in their rotor slots to follow the casing bore, contact with the casing bore generally being maintained by centrifugal force.

Critical clearances exist at the vane ends, the rotor ends, and the vane flanks. Wear is most likely at points of rubbing contact, i.e., the vane tips and flanks. Because there is internal rubbing contact, sliding vane pumps are Group 1 for liquid types.

Conventional sliding vane pumps have a single shaft seal. The rotor is either between bearings, with internal bearings, or cantilever, with external antifriction bearings. For corrosive service or where quick dismantling is required, cantilever rotor construction is usual. Wright [7.2] describes such a line of small sliding vane pumps.

Sealless sliding vane pumps are available, see Wright [7.2] and Capuder et al. [7.3], but only in small sizes to date. The arrangement used is magnetic drive; see Chapter 6 for details of the drive.

7.14. Gear

Two forms of gear pumps are in general use: internal gear and external gear. The pumping action of both is essentially the same; the "displacement volumes" are formed by the spaces between gear teeth and the casing, and displacement is produced by gear meshing. Figure 7.17 shows an internal gear pump, and Fig. 7.18 shows an external gear pump.

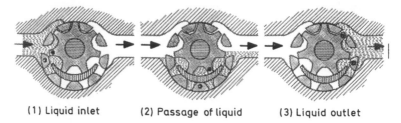

(1) Liquid inlet   (2) Passage of liquid   (3) Liquid outlet

FIG. 7.17.   Internal gear pump.

Gear teeth, in external gear pumps, are either straight spur or herringbone (gapless double helical). Lower flow pulsation and noise are cited as justification for the more expensive herringbone gear teeth.

Gear pump "slip," hence delivered capacity, is dependent on the clearances at the gear ends and gear tips, with the former the more important; see Trushko et al. [7.4]. Torque is split between the gears. Internal gear pumps are Group 1 for liquid types. External gear pumps can be either Group 1 or, with external bearings and gear timing, Group 2.

Harvest [7.5] cites less shear of the pumped liquid as the justification for internal gear pumps. Noting that equivalent external gear pumps are less expensive to manufacture, the use of internal gear pumps would normally be limited to pumping clean, shear-sensitive liquids.

Internal gear pumps have one shaft seal. The driving gear (rotor) can be supported by bearings in the pumped liquid or by external bearings. The driven gear (idler) is supported by a bearing in the pumped liquid.

The simplest external gear pump, Fig. 7.19, has one shaft seal and bearings in the pumped liquid. For more severe service, those handling Group 2 liquids (nonlubricating), external gear pumps are made with external bearings and timing gears and, of necessity, four shaft seals, see Fig. 7.20.

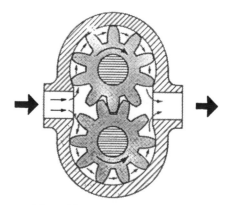

FIG. 7.18.   External gear pump.

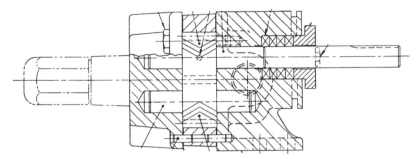

FIG. 7.19. Simple external gear pump; bearings in the pumped liquid, one shaft seal. (Courtesy Worthington Pump, Dresser Industries, Inc.)

When corrosion is not a problem, gear pumps are furnished with iron casing, end plates and gears, steel shafts, and either antifriction or sleeve bearings. Sleeve bearing materials are usually bronze, iron (in alkaline liquids), or carbon, often with antimony, for low viscosity liquids. Moderately corrosive conditions are met with pumps of 316 stainless steel. This alloy is noted for its tendency to gall, thus its success in gear pumps is dependent upon proprietary treatments to reduce galling; see Harvest [7.5]. Taylor [7.6] notes the availability of small gear pumps in alloys higher than 316.

Given the simplicity of the basic external gear pump, Fig. 7.19, there has always been a keen desire to use it in place of the more complex externally timed version. Progress has been made on this. To retain the simplicity and avoid any stagnant regions, most the effort has been directed toward materials. Trushko et al. [7.4] determined that in pumps so equipped, sleeve bearing wear governed pump life. Carbon on ceramic or ceramic on ceramic (aluminum oxide on tungsten carbide or chrome oxide; see Harvest [7.5]) and

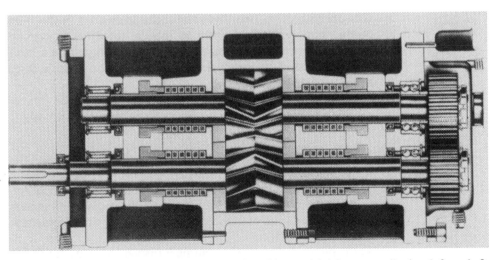

FIG. 7.20. External gear pump for liquids with poor lubricity; externally timed, four shaft seals, external bearings. (Courtesy Worthington Pump, Dresser Industries, Inc.)

PTFE (in small pumps; see Ref. 7.7) are reported as successful bearing materials.

End clearance wear follows bearing wear as the determinant of pump life. Hardened iron (treatment with nonmetallic carbide, nitride, and sulfide; see Ref. 7.7) is one approach. Nonmetallic "wear plates" is another, nonmetallic gears a third.

Gear failure occurs in one of three ways: end face wear, tooth breakage due inadequate strength, or tooth breakage following gross wear. Nonmetallic gears tend to alleviate the first, but strength can be a problem. Reference 7.8 reports success with carbon fiber reinforced polyphenylene sulfide after breakage problems with glass fiber reinforced fluoropolymer. Koster [7.9] notes that the Hertzian stresses in gear pump teeth can be very high (100,000 to 150,000 lb/in.^{2} in typical motive power pumps), and that tooth survival is dependent upon elastohydrodynamic lubrication (EHL). Liquid viscosity, its pressure–viscosity characteristic, and the surface finish of the parts all influence EHL. The significance of this for chemical processing is a need for caution in application lest rapid tooth wear be a problem.

For zero leakage, magnetic drive external gear pumps are available. Such pumps are, of course, essentially for Group 1 liquids, subject to the comments above. As is the case with sliding vane pumps, sizes are presently limited. Nasr [6.33] attributes the limitation to the higher torque required by gear pumps.

When the liquid pumped requires heating in the pump, larger size casings can be furnished with an integral jacket. Smaller pumps are available with bolt-on jackets; see Ref. 7.6.

7.15. Screw Pumps

Designs employing one, two, and three screws are in use. The single screw pump, more generally known as "helical rotor" or "progressive cavity," is really a different class of pumps (Group 3 for liquid type; see Fig. 7.15) and will be discussed separately. In that sense, the pumps to be discussed here are actually "multiple screw."

Following on the fundamental operating principles set out at the beginning of this chapter, multiple screw pumps operate as follows:

The "displacement volume" is opened at the suction as the counter-rotating screws unmesh.

Subsequent meshing of the screws produces a "displacement volume" bounded by the thread flanks, the thread roots, and the pump casing.

Continued rotation of the screws translates the "displacement volume" to the pump discharge.

At the pump discharge the volume is displaced by meshing of the screw ends.

Figure 7.21 illustrates the idea.

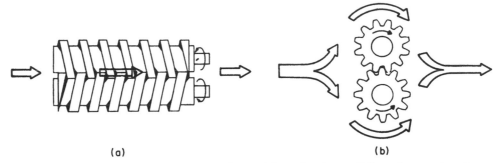

FIG. 7.21. Screw pump translates flow axially rather than radially as in an external gear pump.

Leakage, or "slip," in screw pumps flows to succeeding "displacement volumes." Since displacement volume geometry is essentially identical, and neglecting any changes in liquid characteristic through the pump, the pressure rise across the pump (purely a function of system resistance) will tend to that which balances leakage between the "displacement volumes." The actual pressure rise distribution at any instant is close to equi-stepped, see Fig. 7.22. This characteristic of screw pumps leads to the concept of so much pressure rise per "closure" or "stage." Brennan et al. in Ref. 1.1 cite typical values of 125 to 150 lb/in.^{2} with normal running clearances, and up to 500 lb/in.^{2} with minimum clearances. The import is that the length of the pump is somewhat dependent upon the required pressure rise. Zalis [7.10] draws attention to the need for at least two screw turns to avoid "erratic performance and excessive slip."

A by-product of the gradual pressure build up, one coupled with relatively low flow pulsations and low internal liquid velocities, is low noise. Wegener et al. [7.11] report a "medium size" screw pump having a sound level of 57 dB(A) while running at 2900 r/min and 1450 lb/in.^{2} gauge discharge pressure.

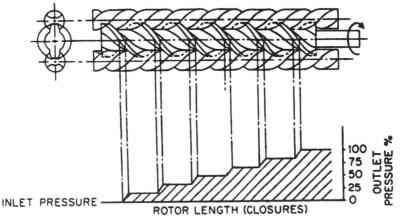

INLET PRESSURE

ROTOR LENGTH (CLOSURES)

OUTLET PRESSURE %

100
75
50
25
0

FIG. 7.22. Pressure rise along a screw pump.

Multiple screw pumps are available in a variety of configurations. The broad distinctions between them are:

Screw rotation: timed or untimed
Construction: single or double end
Number of screws: 2 or 3

In a pump with timed screws, a set of close backlash timing gears controls the rotation of the screws relative to one another, while preventing contact at the thread flanks. With untimed screws the drive screw controls the rotation of those in mesh with it by torque transmitted through the thread flanks.

When the flow direction is axial, the simplest basic pump construction is single end, with the flow passing in one direction through the pumping screws, see Fig. 7.23. The drawback is that the pressure difference across the screws produces axial thrust which must either be absorbed by a thrust bearing or balanced hydraulically. Double end construction, Fig. 7.24, is essentially two opposing pumps operating in parallel. Being opposed, the thrust produced by each "half" pump is balanced. Splitting the pump flow has a further advantage in that it increases the effective suction flow area. Beyond a more complicated casing, double end construction dictates a longer rotor for the same pressure differential.

Two-screw pumps are almost invariably timed. Since each screw does half the pumping, driving the second screw through the thread flanks would entail high losses and produce rapid wear. The timing gears can be internal for clean, lubricating liquids (Group 1) or external for nonlubricating liquids (Group 2); see Figs. 7.25 and 7.26, respectively. Two-screw pumps develop substantial radial thrust, thus require four bearings to support the screws. The bearings are usually antifriction, with their location generally determined by that of the timing gears. When the gears and bearings are internal, the pump has only one shaft seal; when they're external, four shaft seals are required.

Three-screw pumps are not timed. In this arrangement, see Fig. 7.27, the power screw is sealed by two "idler" or "seal" screws. Idler screw rotation is caused by the power screw, but since the only torque transmitted is that necessary to overcome friction, per Wegener [7.11], thread flank wear is not a major consideration. With symmetrical idler screws there is no net radial thrust on the power screw, thus no deflection to allow higher leakage. The idler screws are supported over their entire length by the casing bore. Brennan

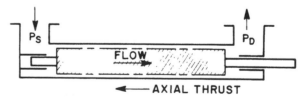

FIG. 7.23. Screw pump: single end construction.

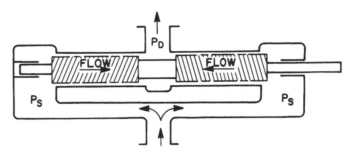

FIG. 7.24. Screw pump: double end construction.

FIG. 7.25. Internally timed, twin-screw pump. (Courtesy Worthington Pump, Dresser Industries, Inc.)

FIG. 7.26. Externally timed, external bearing twin-screw pump. (Courtesy Worthington Pump, Dresser Industries, Inc.)

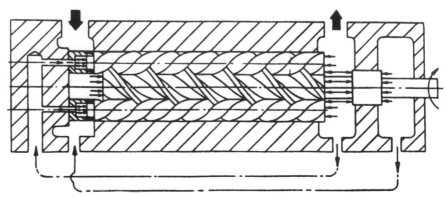

FIG. 7.27. Three-screw pump.

et al. in Ref. 1.1 note that the support relies upon hydrodynamic theory. For this reason, three-screw pumps are only suitable for Group 1 liquid types (clean, good lubricity). Three-screw pumps generally have only one bearing, either internal or external, adjacent to the coupling. With only one shaft passing through the casing, there is only one shaft seal.

Both two- and three-screw pumps are available in either single or double end construction. As a general rule, though, two-screw pumps are furnished double end, three-screw pumps, single end.

Figures 7.24 and 25 show typical double end two-screw pumps. The usual direction of pumping is from end to center, as in Fig. 7.24, since it affords a shorter rotor and maintains the shaft seals at suction pressure. Pumping center to end is used occasionally for liquids of very high viscosity which will not flow easily to each end of the pump. An oversize suction region is required, and it has to be recognized that the shaft seals will be at discharge pressure.

A typical single end, three-screw pump is shown in Fig. 7.27. Suction is generally at the nondrive end since this affords the greatest flow area to the screws. Figure 7.27 also shows how screw thrust is hydraulically balanced. Note that balancing the drive screw also serves to lower the pressure at the shaft seal to approximately suction pressure.

Shaft seals in screw pumps are either packed box or mechanical seal; see Chapter 11.

Screw pump casing construction can be either lined or unlined. In lined construction, see Fig. 7.27, the screws run in a liner located within the casing. Lined construction is nominally more expensive (extra locating fits) but enables better material choices, can facilitate precision machining of the critical bores, and allows their ready replacement. With unlined construction the screws run directly in the casing, see Figs. 7.25 and 7.26. Unlined construction allows lower manufacturing cost and is serviceable where the rate of bore wear is low. When necessary, the casing can be jacketed to permit heating of the liquid as it passes through the pump.

Three-screw pumps are generally lined construction. The usual casing materials are iron, ductile iron, and carbon steel, in order of pressure containment capability. Liners may be iron, treated iron (bore surfaces treated for improved wear resistance), or a soft material such as aluminum. The virtue of a soft liner is a certain capacity to "heal" after passing abrasive solids; see Arcaro [7.12]. Screws have traditionally been a hard material such as nitrided steel. Molded polymers, offering lower cost and adequate performance, are now gaining acceptance.

Two-screw pumps are furnished in both lined and unlined construction. Casing materials include those for three-screw pumps plus more corrosion-resistant alloys such as bronze, aluminum bronze, 316 stainless steel, and higher alloys. Liners, when used, are furnished in a similar range of materials. Screws materials are determined by strength and corrosion resistance, the usual choices being carbon steel, chrome steel, 316 stainless steel, and higher alloys. When the liquid dictates a bore/screw material combination prone to adhesive wear, bore surface treatments (e.g., hard chrome), and screw tip coatings (e.g., Stellite), are employed to realize acceptable service life. Javia [7.13] presented a tabulation of various bore/screw material combinations and their relative performance.

At this stage of their development, screw pumps are not available in sealless configurations.

7.16. Lobe Pumps

Figure 7.28 shows a three-lobe pump and the product flow through it. In principle the lobe pump is similar to the external gear pump; liquid flows into the region created as the counter-rotating lobes "unmesh," "displacement volumes" are formed between the surfaces of each lobe and the casing, and the liquid is displaced by "meshing" of the lobes. The detail differences are two. First, the lobe forms in use, one, two, and three lobe with one rare, are not capable of driving each other, so must be "timed" with separate gears. The lobe pumps produced for chemical processing have their timing gears removed from the pumped liquid [7.14, 7.15], and thus are externally timed. Second, the relatively large "displacement volumes" enable large solids (nonabrasive) to be handled. They also tend to keep liquid velocities and shear low, making the pump type suitable for high viscosity, shear-sensitive liquids.

FIG. 7.28. Rotary lobe pump operation. (Courtesy SSP Pumps Ltd.)

FIG. 7.29.   Overhung rotary lobe pump. (Courtesy SSP Pumps Ltd.)

The choice of two- or three-lobe rotors depends upon solids size, liquid viscosity, and tolerance of flow pulsation. Two lobe handles larger solids and higher viscosity but pulsates more (see System Corrections later in this volume).

With timed rotors there is no contact between them, which confers advantages in rotor material selection, nature of liquid pumped, and inadvertent dry running. Critical clearances are between the lobes themselves, the lobe ends and end plates, and the lobe tips and casing bore.

Small pumps employ cantilever rotors, which simplifies the pump, enables sanitary construction, and requires only two shaft seals, see Fig. 7.29. Larger pumps have simply supported rotors, see Fig. 7.30, and require four shaft seals. Walrath [7.16] observes that larger lobe pumps cost 4–5 times a centrifugal pump of equal flow and head.

Type 316 stainless steel is the "standard" material for wetted parts. Wilson [7.15] notes the availability of alternative rotor materials such as acrylonitrile rubber coated, white nylon, filled PTFE, and bronze when service conditions require them.

Jacketed casings are available if the pumped liquid requires heating.

7.17. Flexible Vane Pumps

Instead of the vanes sliding in response to the profiled casing bore to displace liquid, those in flexible vane pumps deflect; see Fig. 7.31. The advantages of this are a simpler, lower cost pump with some tolerance of abrasive solids in

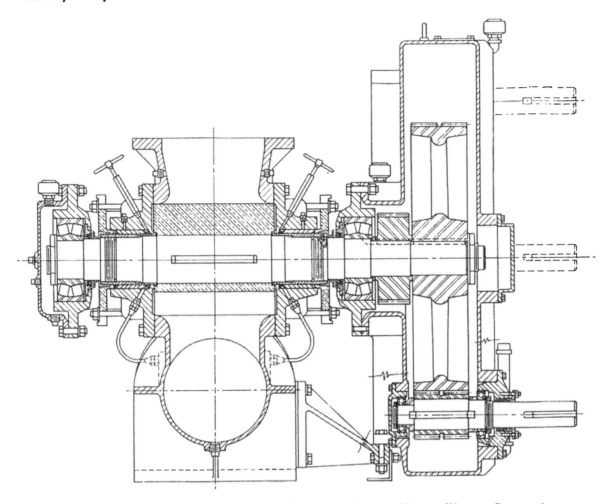

FIG. 7.30. Between bearings rotary lobe pump. (Courtesy Waterous Company.)

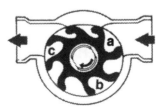

FIG. 7.31. Flexible vane pump operation. (Courtesy ITT Jabsco.)

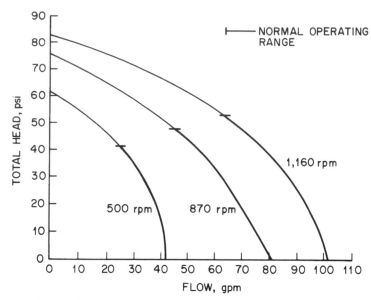

FIG. 7.32. Typical flexible vane pump performance curve [1.2].

the pumped liquid. The disadvantages are limited size and pressure, related to vane stiffness, and limited temperature, a function of vane material.

A further consequence of the flexible vane, a disadvantage in some circumstances but an advantage in others, is a nonlinear pressure versus flow characteristic. Figure 7.32 shows a typical characteristic given by McLean in Ref. 1.2.

Figure 7.33 shows typical flexible vane pump construction. The casing is easily dismantled to permit quick replacement of wearing parts. A cantilever shaft running in two antifriction bearings supports the rotor. A single shaft seal is necessary.

Materials of construction are similar to those for other rotary pumps. Casings are available in iron, bronze, 316 stainless steel, and reinforced plastic. McLean, in Ref. 1.2 identifies glass-reinforced epoxy and phenolic as suitable plastics of corrosion resistance equal to 316 stainless steel. For abrasive service, iron or 316 casings are preferred. The profiled portion of the bore is a replaceable cam made of either plastic or the same material as the casing. Impellers are elastomer, neoprene and nitrite being usual, fluorelasto-mers, polyurethane, and natural rubber, special. Shafts are always metal and usually an alloy whose corrosion resistance is equal to that of the casing.

7.18. Helical Rotor Pumps

Known also as "progressive cavity" pumps, helical rotor pumps are single screw, positive displacement. Being positive displacement, they differ from other single screw pumps which are viscous drag.

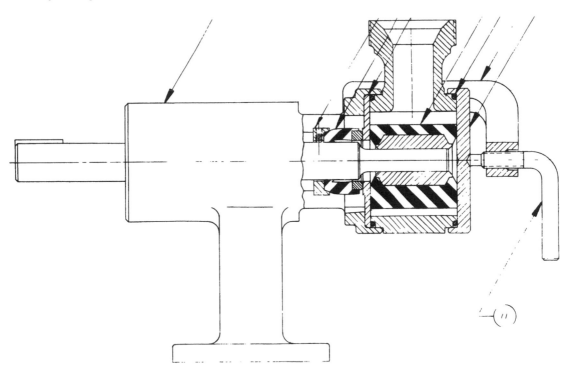

FIG. 7.33. Flexible vane pump section. (Courtesy ITT Jabsco.)

Positive displacement is realized by running a single helical rotor of circular section within a double threaded helix whose pitch is twice that of the rotor. The rotor maintains a seal with the stator in all angular positions, thus forming "displacement volumes" which translate from suction to discharge. The radial position of the rotor at any angular position is dictated by contact with the stator, thus its axis of rotation "orbits." To accommodate the rotor's "orbits," the drive connection to it has to be flexible.

The critical clearance is between rotor and stator. To maintain this, the stator is made of a resilient material, and in many instances has an interference fit with the rotor, either by sizing or pressurizing the outside. Combined with a suitable rotor, the resilient stator affords a notable capability for abrasive solids.

A typical pump is shown in Fig. 7.34. Referring to the construction distinctions given in Screw Pumps, the pump in Fig. 7.34 is single end, which is usual. Flow may be in either direction but is generally from the drive end of the pump, i.e. over the drive shaft. Being single ended without hydraulic balancing, the pump's allowable pressure rise is limited by axial thrust on the rotor drive (in compression) and the thrust bearing.

Rotor drive shaft flexibility is afforded by either encapsulated universal joints or a slender high strength "flexible" shaft whose deflection does not impose intolerable loads.

A single shaft seal is required; it usually operates at suction pressure and can be either packed box or mechanical. For clean liquid applications

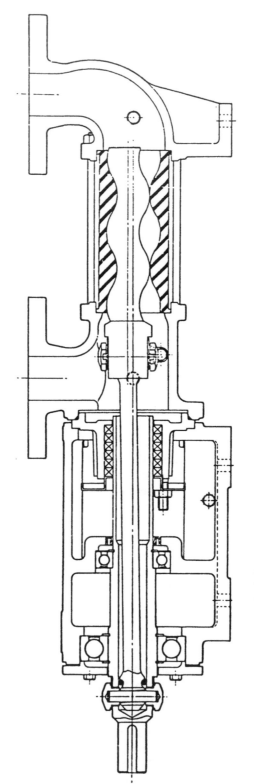

FIG. 7.34. Helical rotor pump. (Courtesy Worthington Pump, Dresser Industries, Inc.)

TABLE 7.4   Rotor and Stator Materials for Various Services

Part	Material	Service
Rotor	Nitrided steel	Abrasive, noncorrosive
	Chrome plated 316	Corrosive, with or without abrasives
	Monel or high nickle-molybdenum alloys	Highly corrosive
Stator	Natural rubber	General
	Nitrile rubber	Oil, fats, effluent
	Cast urethane	Abrasive, aqueous slurries
	Hypalon[a]	Mineral acids, oxidizing chemicals
	Viton[a]	Aliphatic and aromatic hydrocarbons, high temperature

[a]Registered trademark of DuPont.

involving a vacuum at the suction, flow is sometimes reversed to be toward the drive so the seal is at discharge pressure.

Power is transmitted from the driver via a drive shaft. Two antifriction bearings support the drive shaft, and one of them also carries the rotor's hydraulic axial thrust. In some arrangements either bearing can act as the thrust bearing.

Casing materials range from iron through 316 stainless steel to higher alloys, depending upon corrosiveness and pressure. For the rotor and resilient stator, Goodchild [7.14] suggests the information in Table 7.4.

"Flexible" drive shafts must remain corrosion-free to realize their design endurance strength. Wilson [7.15] notes the use of "impermeable polymeric" coatings to achieve this.

For high viscosity liquids, where gravity flow into the pump is a problem, helical rotor pumps are available in the so-called "open throat" configuration, see Fig. 7.35. In this arrangement the suction opening is oversize and an auger serves to ensure that each of the pump's cavities is completely filled.

As is the case for multiple screw pumps, helical rotor pumps are not currently available in sealless form. Some work has been done on wet pit versions, but pumps so constructed are not yet in regular service.

7.19. Peristaltic Pumps

The peristaltic or flexible tube pump is unique in that it's the only sealless rotary pump with solids handling capability. Maynard [7.17] laments that despite this uniqueness, the device is underutilized.

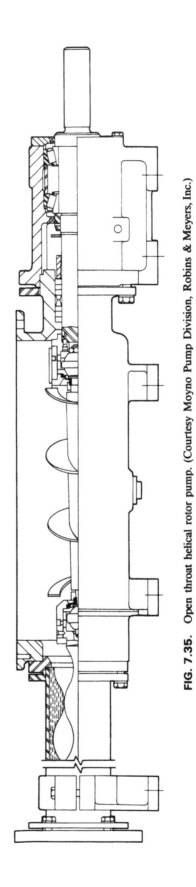

FIG. 7.35. Open throat helical rotor pump. (Courtesy Moyno Pump Division, Robins & Meyers, Inc.)

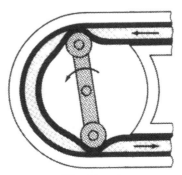

Flexible tube pump.

FIG. 7.36. Flexible tube (peristaltic) pump. (Reprinted with permission from Ref. 1.1.)

Figure 7.36 illustrates the principle; a "displacement volume" is captured ("occluded") by the rotor closing off the tube at the suction, then displaced to the discharge by continued rotation.

Early peristaltic pumping arrangements applied the mechanism to a hose between suction source and discharge point. Doing this, termed a nondedicated hose, avoided any joints in the transfer piping, but at the same time restricted pump performance and, because the hose was not entirely suitable, led to short hose life. Figure 7.37 shows the performance deterioration with decreasing suction pressure or increasing discharge pressure. Poor hose restitution and sealing account for the performance deterioration.

By using a discrete hose, one designed to aid the pumping action and including various refinements for occlusion, performance has been improved. Current design peristaltic pumps exhibit a true positive displacement characteristic, i.e., at a given speed there is no change in flow when suction and discharge pressure is varied over the pump's working range.

Occlusion refinements include three roller rotors and sliding shoes, Figs. 7.38 and 7.39. The former reduces flow pulsation; the latter, hose stress.

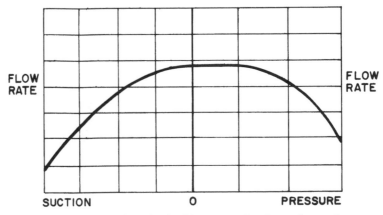

FIG. 7.37. Performance variation of peristaltic pump against decreasing suction pressure and increasing discharge pressure.

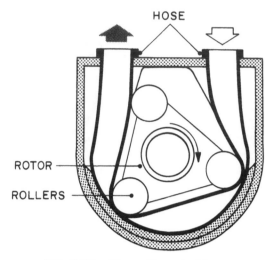

FIG. 7.38.   Three roller peristaltic pump.

The only component in contact with the pumped liquid is the hose, therefore the only variable material is that of the hose. In sophisticated designs, only the inner lining of the hose will change with the liquid pumped; the outer covering and its reinforcement remain the same.

Maynard reports discrete hose lives of 2000–4000 in abrasive chemical process service. Failure is almost exclusively by repeated stress from occlusion, even in services handling large abrasive materials.

FIG. 7.39.   Sliding shoe peristaltic pump. (Courtesy Waukesha Pumps.)

References

7.1. S. E. Dolby and G. Carter, "Helical Rotor Pumps—Viscous Performance," *World Pumps*, May 1985.

7.2. S. E. Wright, "A Philosophy for Pump Design," *World Pumps*, pp. 186–189 (April 1982).

7.3. F. C. Capuder et al., "Recent Developments in Magnetically Coupled Vane Pumps for Tritium Service," *Fusion Technol.*, *8*, 2420–2422 (September 1985).

7.4. P. V. Trushko et al., "Increasing the Durability of Gear Pumps," *Sov. Eng. Res.*, *2*(6), 9–11 (January 1982).

7.5. J. Harvest, "Recent Developments in Gear Pumps," *The Chem. Eng.*, *403*, 28–29 (May 1984).

7.6. J. M. Taylor, "Small Gear Pumps," *Plant Eng.*, *32*(23), 177–180 (November 9, 1978).

7.7. Anonymous, "A Pump That Lasts Longer Than Most," *Ind. Lub. Tribol.*, *32*(3), 94–95 (May/June 1980).

7.8. Anonymous, "Carbon Fiber-Reinforced PPS Prolongs Gear Life." *Des. News*, p. 108 (November 23, 1987).

7.9. T. P. Koster, "The Selection and Application of Gear Pumps and Motors for Fire-Resistant Fluids," *CIM Bull.*, *73*(823), 89–95 (November 1980).

7.10. A. A. Zalis, "Rotary Screw Pumps," *Plant Eng.*, *33*(23), 197–201 (November 15, 1979).

7.11. J. Wegener et al., "Screw Pumps of One, Two, and Three Screw Design," in *Proceedings of the 2nd International*, Pump Symposium, Houston, Texas, April 30–May 2, 1985, pp. 41–46.

7.12. R. J. Arcaro, "The Screw Pump in Fuel Service," *Power*, *123*(7), 97–102 (1979).

7.13. R. H. Javia, "Designing Twin Screw Pumps," *Adhes. Age*, pp. 35–38 (February 1982).

7.14. B. Goodchild, "Line Transfer of Corrosive Materials," *CME*, *28*(7), 34–36 (1981).

7.15. R. D. Wilson, "Special Pumps for Special Applications," *Processing*, *21*(3), 44–46 (1975).

7.16. D. Walrath, "Pumps and Driver Selection," in *Conference Proceedings of Pumping Station Design for the Practicing Engineer*, *#III*, *Wastewater*, Bozeman, Montana, 1981, Chapter 2.4, Section E.

7.17. R. Y. Maynard, "Modern Developments in Peristaltic Pumping," *Pumps and Pumping*, *A Practical Guide to Recent Developments*, Institute of Chemical Engineers, Manchester, UK, November 1985, pp. 7.1–7.14.

8. Reciprocating Pumps

8.1. Fundamental Operation

Instead of moving their "displacement volumes" from suction to discharge, as do rotary pumps, reciprocating pumps create and displace their "displacement volumes" by the action of a reciprocating element. Suction and discharge flow directions are controlled by valves. Figure 8.1 illustrates the principle.

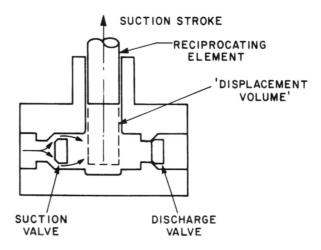

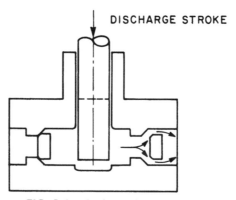

FIG. 8.1.   Reciprocating pump operating principle.

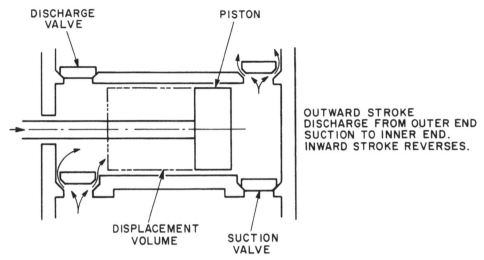

FIG. 8.2. Double acting reciprocating pump.

8.2. Basic Construction

Four forms of reciprocating element are in use: piston, plunger, (Fig. 8.1), diaphragm, and what can be termed "mobile separator." Details of each form are given in specific sections later in this chapter.

Plungers, diaphragms, and mobile separators are single acting, i.e., each element can discharge in one direction only. Pistons can be either single or double acting, Fig. 8.2 illustrating the latter arrangement.

Reciprocating pumps comprise two major segments: a liquid end and a power end. Figure 8.3 shows the distinction for a typical vertical plunger pump. Liquid end types are classified by the type of reciprocating element and its action where applicable. Power ends provide the reciprocating motion and can be either slider-crank (known as a power pump) or direct acting. Figure 8.3 shows a vertical power pump, and Fig. 8.4 shows a classical direct acting steam driven piston pump. See Sections 8.17 and 8.18 for detailed descriptions.

8.3. Flow Characteristic

By their very nature, reciprocating pumps produce pulsating flow. A single acting, simplex pump would produce flow for only 1/2 of each cycle; Fig. 8.5(a). Making the pump double acting, Fig. 8.5(b), gives two pumping strokes per cycle, but the flow still stops and starts at the end of each stroke. Adding a second cylinder to make the pump duplex, and timing the pumping strokes for a small overlap, Fig. 8.5(c), significantly reduces flow pulsation.

FIG. 8.3. Vertical plunger pump section. (Courtesy Worthington Pump, Dresser Industries, Inc.)

FIG. 8.4. Direct acting piston pump. (Courtesy Worthington Pump, Dresser Industries, Inc.)

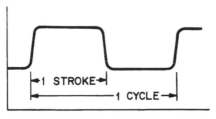

(a) SINGLE ACTING, SIMPLEX

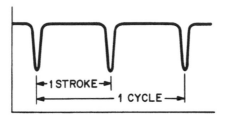

(b) DOUBLE ACTING, SIMPLEX
SINGLE ACTING, DUPLEX

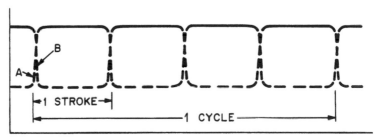

(c) DOUBLE ACTING DUPLEX

FIG. 8.5. Direct acting pump flow.

For usual designs of power pumps, the reciprocating motion approaches sinusoidal, giving a duplex flow characteristic of the form shown in Fig. 8.6(a). By using a multiple, odd number of cylinders to overlap the individual cylinder flows, the pump's flow pulsation can be reduced. Figures 8.6(b) and 8.6(c) show the resultant flow characteristics for single acting triplex and quintuplex pumps. The usual values of flow pulsation are shown in Table 8.1.

8.4. Flow versus Pressure

For a given pump, speed, and liquid, pump mean flow is always lower than displacement and decreases slightly with increasing pressure, see Fig. 8.7. The

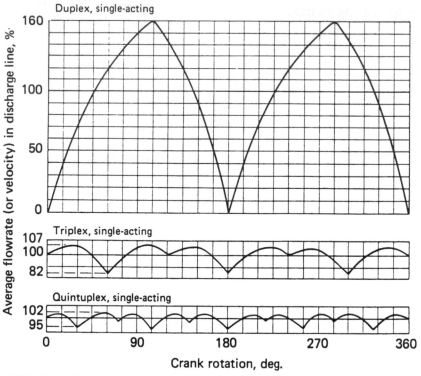

FIG. 8.6. Flow curves for reciprocating power pumps. (Courtesy Terry L. Henshaw.)

TABLE 8.1 Flow Pulsation for Reciprocating Pumps

Number of Single Acting Pump Elements	Pump Type	Flow Pulsation (%)
2	Duplex	160
3	Triplex	25
5	Quintuplex	7
7	Septuplex	4
9	Nonuplex	2

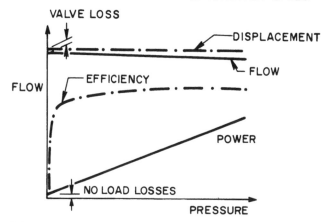

FIG. 8.7. Typical reciprocating pump performance characteristic.

ratio of pump flow (by convention suction flow; Henshaw in Ref. 1.3) to displacement is termed volumetric efficiency, i.e.,

$$\text{Volumetric efficiency} = \frac{\text{pump flow at suction}}{\text{pump displacement}} \qquad (8.1)$$

Volumetric efficiency has two components: slip and compressibility.

Slip is comprised of valve and seal (rod, piston, or plunger) leakage. Valve leakage increases with speed and pressure. Seal leakage is low (considered negligible in power pumps; Buse in Ref. 1.1) and essentially dependent upon pressure.

Compressibility reduces pump capacity by retarding suction valve opening, as the liquid remaining in the "clearance volume" has first to expand to suction pressure. The magnitude of capacity reduction due to compressibility depends upon liquid end geometry (ratio of total volume to displacement) and liquid bulk modulus of elasticity.

Pump sizing, i.e., displacement, has to be corrected for volumetric efficiency. The process is iterative because the correction is influenced by the pump selection.

Direct acting pumps cannot develop a pressure greater than that of the motive fluid times the ratio of drive to pump effective areas. Increasing the difference between motive fluid and pump pressures raises the force available to move the liquid, hence the pumping rate. Figure 8.8 shows the resultant characteristic for a direct acting pneumatic diaphragm pump.

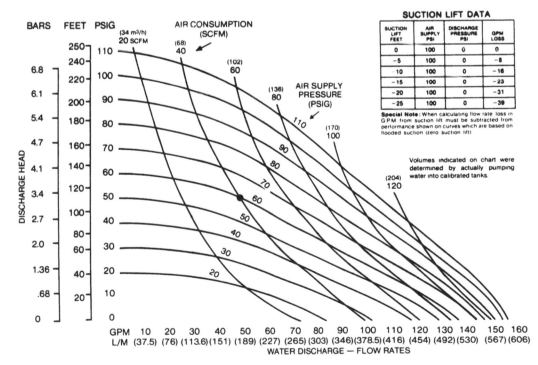

FIG. 8.8. Performance characteristic of a direct acting reciprocating pump. (Courtesy Wilden Pump & Engineering Co.)

8.5. Power

At constant speed, power varies nearly linearly with differential pressure, see Fig. 8.7. The "no load" power is the sum of mechanical losses (friction) and liquid end hydraulic losses for that flow. Above a certain minimum speed (necessary for bearing lubrication), power pump mechanical losses are low and increase only slightly with speed. Direct acting pumps incur higher friction losses, which increase significantly with speed.

Pump efficiency, often termed "mechanical efficiency" is defined as

$$\text{Pump efficiency} = \frac{\text{hydraulic power}}{\text{power absorbed}} = \frac{Q(\Delta P)}{1714(\text{hp})} \tag{8.2}$$

This efficiency covers the losses incurred by mechanical friction, slip (or leakage), and the nonrecovered portion of liquid compression. Because volumetric efficiency is partially an effectiveness, i.e., part does not incur an energy loss, pump and volumetric efficiency, while related, are not tied to one another.

Power plunger pump efficiency for usual suction pressures is typically 90%. Power piston pumps 90% for double acting, 85% for single acting. Direct acting pump efficiency is affected by stroke length as well as speed. Freeborough, in Ref. 1.1, gives values ranging from 50 through 75% for steam driven piston pumps of 3 through 24 in stroke running at 50 ft/min.

Some power pump applications involve high suction pressures with the differential pressure being significantly less than discharge. The hydraulic power is calculated for the suction flow and pump differential pressure. Pump power, however, has to be determined from the power to develop the discharge pressure less the power recovered from the suction:

$$\text{Net hp} = \frac{QP_d}{1714(\eta)} - \frac{QP_s(\eta')}{1714} \tag{8.3}$$

where P_d = discharge pressure, P_s = suction pressure, η = pump efficiency, and η' = efficiency of the pump as a motor (3–5 points lower than pump efficiency; Henshaw in Ref. 1.3). Because the power recovery efficiency is lower, the net pump power is higher than usual, thus the efficiency is lower.

8.6. NPSHR

Reciprocating pump NPSHR is currently defined as the NPSHA at 3% capacity reduction. Proposals now under consideration will extend the definition to include NPSHA at an audible indication of cavitation. Figure 8.9 is a typical test plot showing flow versus NPSHA with the pump running at constant speed.

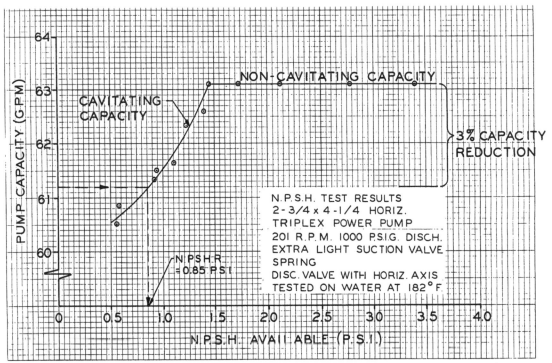

FIG. 8.9. Reciprocating pump capacity versus NPSH available. (Courtesy Terry L. Henshaw.)

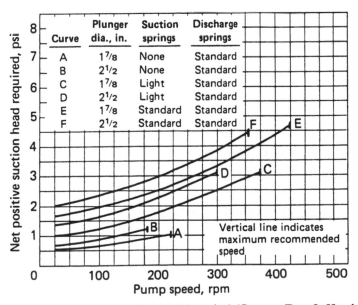

FIG. 8.10. Valve loading affects NPSH required. (Courtesy Terry L. Henshaw.)

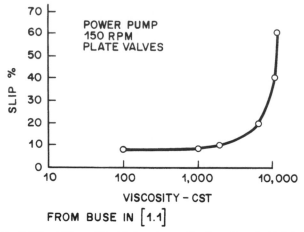

FIG. 8.11.   Effect of liquid viscosity on slip (from Buse in Ref. 1.1).

The principal component of reciprocating pump NPSHR is the pressure differential required to open the suction valve. NPSHR is therefore expressed in pressure terms and increases with pump speed and valve closing force. Figure 8.10 shows a typical range of NPSHR characteristics.

With NPSHR defined by a specific deterioration in performance, evidence of substantial cavitation, there arises the question of damage when running with NPSHA equal to or just above NPSHR. Damage has occurred under these conditions, but so far there are little data to indicate what margin over NPSHR will insure no damage. Buse, in Ref. 1.1, suggests a margin of 3 to 5 lb/in.^{2} based on "good practice."

8.7. Liquid Viscosity

Increasing viscosity has two effects. First, the motion of the valves is impeded. At constant speed this leads to greater valve leakage, hence lower volumetric efficiency, see Fig. 8.11. Alternatively, for a given volumetric efficiency, pump speed must be reduced as viscosity increases, see Fig. 8.12. The second effect is increased head loss through the suction valve and port, resulting in higher NPSHR. The extent of this effect increases with pump design pressure.

8.8. Entrained Gas

Reciprocating pumps will compress gas, but the pressure developed is limited by the usually large clearance volume. Unless the pump can discharge some gas during each stroke, it will not clear itself. A second and potentially more

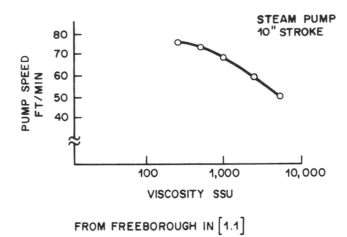

FROM FREEBOROUGH IN $\left[1.1\right]$

FIG. 8.12. Effect of liquid viscosity on pump speed (from Freeborough in Ref. 1.1).

serious problem arises if the pump is able to raise the pressure sufficiently to have the gas go into liquid solution. Should this happen, the sudden change in volume can produce destructive pressure pulsations.

8.9. Interaction with System

At constant speed, handling a given liquid, reciprocating pumps deliver essentially constant flow regardless of system resistance. Volumetric efficiency does fall with increasing differential pressure, but not to the same extent as rotary pumps.

Direct acting pumps with motive fluid at a fixed pressure exhibit a reduction in flow with increasing system resistance, finally reaching a point where the pump stalls.

In both cases the NPSHA must be greater than NPSHR for the pump to deliver its rated flow. With NPSHA below NPSHR, the pump's capacity will be reduced. In severe cases the shocks associated with vapor collapse (a sudden reduction in pumped fluid volume) can cause fracture of major pump parts.

Figure 8.13 illustrates the system interaction for both forms of reciprocating pump.

8.10. Parallel Operation

Given the flow characteristic of constant speed reciprocating pumps, their operation in parallel is quite straightforward; flows are cumulative regardless

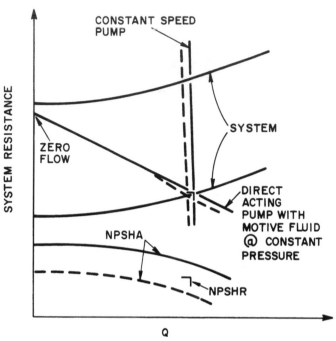

FIG. 8.13. Interaction between reciprocating pump and system.

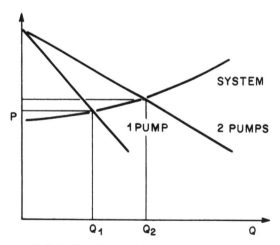

FIG. 8.14. Direct acting pumps in parallel.

of pump size or system resistance. The only caution is the need to account for possible synchronization of flow pulsations when calculating NPSHA in systems with a common suction line.

Direct acting pumps, those with a characteristic such as is shown in Fig. 8.8, will operate in parallel. Flows will be cumulative for the same pressure, see Fig. 8.14. When the system resistance has a friction component, two equal size pumps in parallel will not double the flow because the operating pressure is higher.

In some instances, a small increase in capacity in high pressure service, for example, it may be desirable to operate a centrifugal and reciprocating pump in parallel. Figure 8.15 shows the characteristics of such an arrangement. There are two concerns. First, the centrifugal pump has to develop sufficient head to ensure it operates at an acceptable flow at the higher system resistance. Second, unless the system includes an effective pulsation dampener or is entirely static, the centrifugal pump's flow will fluctuate in concert with that of the reciprocating pump. Centrifugal pumps are usually not able to tolerate continual fluctuations in torque.

8.11. Series Operation

Given their basic flow characteristic, there is little justification for operating reciprocating pumps in series. Driver size and pipeline pressure are two possible reasons. When series operation is contemplated, great care is necessary. The problem is similar to but more severe than that in rotary pumps: small differences in pump geometry or speed can lead to significant pressure differences. To avoid difficulty, series pump installations must be designed recognizing that the first pump determines the flow. Successive pumps must then be provided with automatic flow control, responding to suction pressure, to ensure the total pressure rise is shared. Some capacitance between each pump is advisable to accommodate control lag.

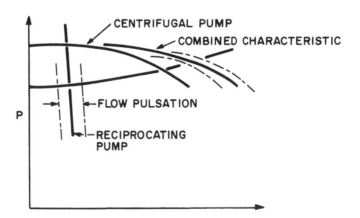

FIG. 8.15. Reciprocating pump in parallel with a centrifugal pump.

Suction boosting reciprocating pumps with a centrifugal pump to provide NPSH is a common practise. The operating flow range must be within the centrifugal pump's capability, and the flow pulsations must be eliminated with a dampener. Since the centrifugal pump has the lower efficiency, its contribution to the total pressure rise is generally limited to that needed for NPSH.

8.12. Varying Capacity

With direct acting pumps, flow is varied by regulating the motive fluid supply. For power pumps, three means are employed. Two allow a range of flows while the third allows only rated flow or zero flow.

8.12.1. Bypass

Bypassing is the lower capital cost means of varying flow. Surplus flow is bypassed back to the pump's suction vessel, see Fig. 8.16. The control parameter for the bypass valve is typically process flow, process pressure, or, in series pumping, pump suction pressure. With bypass flow control, pump power varies only with discharge pressure, thus any bypassing represents an energy loss. When the differential pressure is high, the energy loss may also cause rapid erosion of the bypass valve. Piping and block valves downstream of the bypass control valve should be rated for pump maximum discharge pressure.

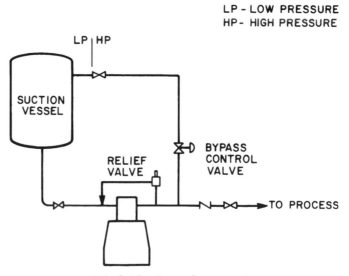

FIG. 8.16. Bypass flow control.

8.12.2. Variable Speed

Varying pump speed to control flow incurs a higher capital cost than bypassing but can represent a substantial energy saving. For electric motor drivers, the most common, there are two usual means of varying pump speed: a slip type device between the pump or gear and motor or modulating power supply frequency to the motor. Figure 8.17 shows the two arrangements diagrammatically. For other than very small reductions in speed, frequency modulation (VFD) absorbs less energy. Combining this with advances in reliability and cost reduction, VFD is now generally the first choice for speed variation. Whether to bypass or vary speed depends upon the extent and duration of operation below rated flow.

8.12.3. Unloading

Pump capacity is reduced to zero by holding the suction valves open during the discharge stroke. Unloading is synchronized with the pump's discharge stroke sequence for a smooth change in flow state. Unloading is suitable only when the process can tolerate a step change from rated to zero flow. Smith, in Ref. 1.1, details unloading mechanisms.

$$KW = \dfrac{MOTOR}{LOSSES} + \dfrac{DRIVE}{LOSSES} + \dfrac{GEAR}{LOSSES} + \dfrac{PUMP}{POWER}$$

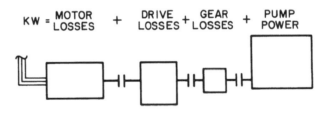

(a) SLIP TYPE DRIVE

$$KW = \dfrac{VFD}{LOSSES} + \dfrac{MOTOR}{LOSSES} + \dfrac{GEAR}{LOSSES} + \dfrac{PUMP}{POWER}$$

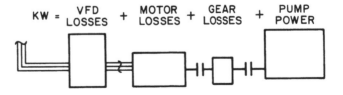

(b) VARIABLE FREQUENCY DRIVE (VFD)

FIG. 8.17. Variable speed drive.

8.13. Tolerance of Solids

With appropriate rating and construction, reciprocating pumps have a high tolerance of solids and are used for those services beyond the capability of Group 3 rotary pumps (see Chapter 7). Appropriate rating and construction means lower speeds, arrangements to exclude abrasives from fine clearances, and wear-resistant materials. Further details are given in the descriptions of each pump type.

8.14. Suction Corrections

Pulsating flow dictates care with both the suction and discharge systems of reciprocating pumps.

At the suction, when there is no pulsation dampener, NPSHA has to be corrected for suction line head loss at peak flow and for acceleration head. Flow variations for various power pump configurations are given in Section 8.3, "Flow Characteristics." Acceleration head can be estimated using the following equation from *The Standards of the Hydraulic Institute* [5.1]:

$$H_a = \frac{LVNC}{kg} \tag{8.4}$$

where H_a = acceleration head, feet of pumped liquid; L = length of suction line (run not equivalent), ft; V = liquid velocity in suction line, ft/s; N = crankshaft speed, r/min; C = constant depending on pump type; k = constant depending on liquid; and g = gravitational constant, 32.2 ft/s². Values of C and k are shown in Table 8.2.

Henshaw, in Ref. 1.3, cites reports suggesting that Eq. (8.4) with the constants listed in Table 8.2 gives conservative results.

TABLE 8.2 Values for C and k for Reciprocating Pumps.

Pump Type	C
Single acting duplex	0.200
Triplex	0.066
Quintuplex	0.040
Septuplex	0.028
Nonuplex	0.022

Liquid	k
Noncompressible, e.g. deaerated water	1.4
Most liquids	1.5
Compressible, e.g., ethane	2.5

Pulsating discharge flow manifests itself as pressure pulsations. Depending upon the flow rate, pipeline size and length, liquid pumped, and the pump type, these pressure pulsations may be sufficient to (1) significantly increase pumping power or (2) produce severe piping vibration. To avoid the latter, Davidson [1.5] suggests pressure pulsations, ΔP, be less than $0.04(P_D)^{0.7}$, where P_D = mean pump discharge pressure. Reducing pressure pulsations involves the installation of pulsation dampeners; see Wachel and Szenosi in Ref. 1.1 and Ekstrum in Ref. 1.3.

Both the suction and discharge systems are susceptible to having accoustic liquid resonance amplify pressure pulsations. Simple piping systems (desirable) can be analyzed manually while complex arrangements require computer analysis; see Wachel and Szenosi in Ref. 1.1.

8.15. Pump Selection

Figure 8.18 sets down a pump-type selection process based first on the nature of the liquid and then on pump pressure capability. The first distinction is

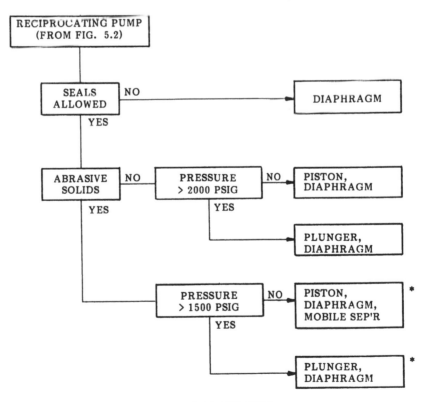

* DESIGNS FOR SOLIDS HANDLING

FIG. 8.18. Reciprocating pump selection.

PUMP TYPE	ACTION	POWER BHP	Q GPM	P PSIG	VISC SSU	TEMP °F
DIAPHRAGM	DIRECT POWER	– 400	250 750	125* 5,000+	50,000	300 400
PISTON	DIRECT POWER	– 3500	10,000	2,000 2,000	5,000 5,000	400 600
MOBILE SEPARATOR	DIRECT	–	–	600	–	–
PLUNGER	DIRECT POWER	– 2500	2500	10,000 10,000	5,000	800

* LIMIT OF USUAL MOTIVE FLUID-COMPRESSED AIR

FIG. 8.19.   Reciprocating pump capabilities.

whether seals are allowed, i.e., whether leakage to the atmosphere can be tolerated. When seals are not allowed, a diaphragm pump is the only choice. Whether the liquid contains abrasive solids is the second distinction. Broadly speaking, the same basic pump types are used for either case, but pumps for solids handling incorporate special design features. Mobile separator pumps are peculiar to solids handling.

Where several types are suitable for the liquid, the choice is made on the basis of capability. Figure 8.18 includes broad pressure capability. Additional data on typical capabilities are tabulated in Fig. 8.19.

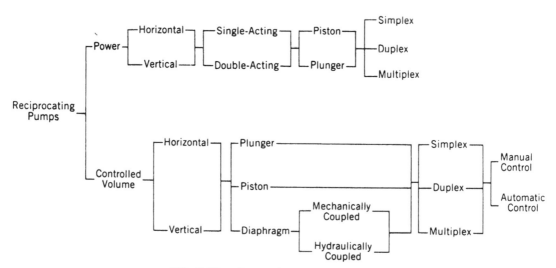

FIG. 8.20.   Classes of reciprocating pumps.

8.16. Reciprocating Pumps

Reciprocating pumps are classified as shown in Fig. 8.20. The essential distinctions are power end and liquid end. Noting this, the description of pump types follows the sequence set out below:

Power end:
    Power pumps
    Direct acting
Liquid end:
    Piston
    Plunger
    Diaphragm
    Mobile separator

8.17. Power Pumps

Power pumps offer high mechanical efficiency and a true positive displacement flow characteristic.

Reciprocating motion is obtained with a slider crank mechanism, see Fig. 8.21. A rotating eccentric, acting through a connection rod, causes the slider to reciprocate. The reciprocating motion is not sinusoidal but approaches it as the ratio of connecting rod length to crank eccentricity, L/R from Fig. 8.21, becomes larger. Usual values for L/R range from 4:1 to 6:1. Lower values produce unacceptably high component inertia forces and liquid acceleration head; higher values produce inordinately expensive power ends.

Power end configurations are distinguished by slider axis orientation, either horizontal or vertical, and discharge stroke direction, either inward or outward. Factors influencing slider axis orientation are shown in Table 8.3.

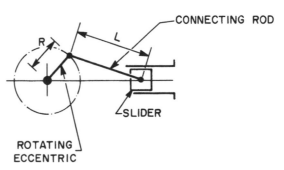

FIG. 8.21. Slider crank mechanism.

TABLE 8.3   Factors Influencing Slider Axis Orientation

	Slider Axis Orientation	
Factor	Horizontal	Vertical
NPSH	Δ	—
Shaking forces	—	Δ
Machine size	Δ	—
Reciprocating element support	—	Δ
Lubrication	Δ	—

TABLE 8.4   Factors Influencing Discharge Stroke Direction

	Discharge Stroke	
Factor	Inward	Outward
Frame and mechanism loading	Δ	—
Machine cost	—	Δ
Seal separation and access	Δ	—

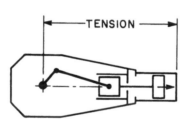

(a) SINGLE ACTING, DISCH STROKE OUT

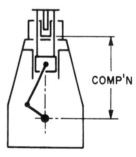

(b) SINGLE ACTING, DISCH STROKE IN

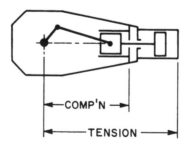

(c) DOUBLE ACTING

FIG. 8.22.   Power pump frame loading.

Factors influencing discharge stroke direction are shown in Table 8.4.

Frame and mechanism loading is a decisive factor and is determined by liquid end design. Figure 8.22 shows the three possible arrangements and their resultant frame loading. Single acting pumps with the discharge stroke outward, Fig. 8.22(a), load the frame in tension and the mechanism in compression. Reversing the discharge stroke direction, Fig. 8.22(b), puts the frame in compression and the mechanism in tension. Double acting pumps, Fig. 8.22(c), load the frame and mechanism alternately in tension and compression.

In most power pumps the "slider" is a separate crosshead, motion being imparted to the pump by rods in tension or compression. With this arrangement, sealing between the power end and liquid end is aided by separation, see Fig. 8.23. Lower cost can be realized by making the "slider" the reciprocating element. The drawbacks are risk of pump seal leakage directly into the power end and imposition of radial loads on the pump seal. Mechanically actuated diaphragm pumps, a form of power pump, are an exception since there is no leakage from the pump.

Out of the factors given, current power pump configurations are generally horizontal piston pumps to 3500 hp, horizontal plunger pumps to 200 hp (some designs to 600 hp), and vertical plunger pumps to 2500 hp.

The change in plunger pump configuration derives primarily from frame loading. Above 200 hp the frame is less complicated if in compression. An inward discharge stroke, Fig. 8.22(b), requires tie rods to the plunger. Installation and operation of this arrangement is easier when it's vertical.

Materials in general use for the major components of power pump power ends are:

Frame	Cast iron, fabricated steel (large, special designs).
Crankshaft	Ductile iron, cast steel, forged steel.
Connecting rod	Tension—cast steel, forged steel.
	Compression—ductile iron, aluminum.
Crosshead	Cast iron
Bearings	Main, crank pin, and wrist pin can be either sleeve or antifriction. Sleeve bearings are either babbitt on bronze or tri-metal, the latter having greater capacity.

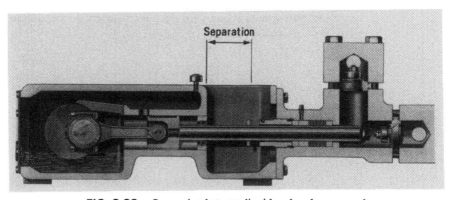

FIG. 8.23. Separation between liquid end and power end.

"Splash" lubrication is used for small, horizontal pumps. When the bearing loading requires it or the pump is vertical, filtered, pressure feed lubrication is used. A heat exchanger is included if necessary.

Power pump power end operating limits are determined by:

Rotative speed. Design speeds range from 800 r/min down to 300 r/min as size increases. Sleeve bearing units cannot run loaded below approximately 40 r/min (lack of hydrodynamic action). Antifriction bearing units can start loaded.

Rod load. Nominally related to crankshaft size, number of main bearings, and size and material of the main, crank pin, and wrist pin bearings.

Suction pressure. High suction pressures in single acting pumps produce unidirectional mechanism loads. These impair the lubrication of sleeve-type wrist pin bearings and crossheads.

8.18. Direct Acting

Although less efficient than power pumps, direct acting power ends afford flexibility in motive fluid and pump regulation. Operation submerged and in hazardous environments are two examples.

The operating principle is straightforward. Motive fluid at some constant pressure acts on an element connected directly to the pump, producing a force greater than the sum of pump resistance plus mechanism friction. Pumping rate is determined by the force margin. When the drive and pump elements are pistons, as in Fig. 8.24, the piston area ratios can be varied to yield the desired pumping rate with a given motive fluid source. A diaphragm, Fig. 8.25, in which power and pump end areas are effectively equal, cannot pump against a pressure greater than that of the motive fluid.

Piston-type power ends obtain reciprocating motion by switching the motive fluid to either side of the piston. Valves whose action is controlled by piston rod position do this, see Fig. 8.4. In duplex pumps the valves are set to provide the small stroke overlap necessary for smooth flow, Fig. 8.5(c). See Freeborough in Ref. 1.1 for details of piston power ends and valving.

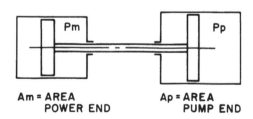

FORCE BALANCE: $Pm\, Am = Pp\, Ap + Ff$

FIG. 8.24.   Force balance in direct acting piston pump.

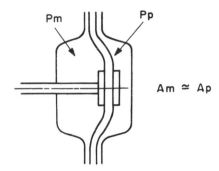

FIG. 8.25. Direct acting diaphragm pump has pump and motive pressures acting on approximately equal areas.

Because the motive fluid can't be switched to the other side of the diaphragm, direct acting diaphragm pumps are duplex with a mechanical connection between the motive fluid sides of their back-to-back diaphragms. Figure 8.26 shows the concept. By switching the motive fluid from one diaphragm to the other, reciprocating motion is obtained. Since the diaphragms are linked, flow must stop from one before it can start from the other, giving a pulse flow characteristic, see Fig. 8.5(a).

Piston-type power ends are generally designed for steam, thus iron construction is standard. Higher steam pressures require ductile iron or steel cylinders, the latter iron lined. Piston and valve rods can be steel, chrome steel, or Monel.

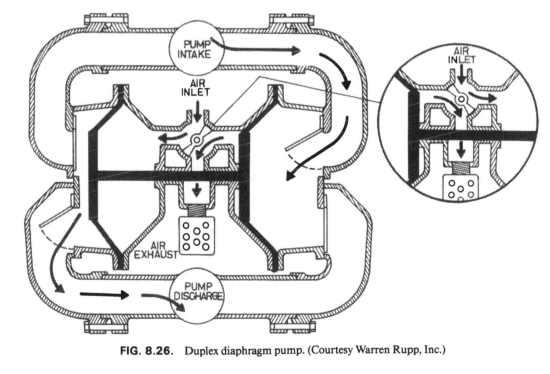

FIG. 8.26. Duplex diaphragm pump. (Courtesy Warren Rupp, Inc.)

The materials of diaphragm type power ends are largely determined by the needs of the liquid end.

Direct acting power ends are limited by:

Speed. Based on mechanical efficiency, the upper limit is approximately 50 cycles/min. There is no lower limit; direct acting pumps can run right down to zero speed, which is sometimes an advantage.

Thermal efficiency. The motive fluid does not expand during the pumping stroke, resulting in low thermal efficiency for the pump. If the exhaust motive fluid can be put to some other use, some of its energy can be recovered. If not, it's all lost in throttling across the exhaust.

8.19. Piston Pumps

A piston is a reciprocating element that incorporates a seal, Fig. 8.27. This arrangement has two advantages: With appropriate valving the pump can be double acting; for a given reciprocating element mass and number of pumping strokes, capacity is greater than with plunger pumps. The limitations are pressure differential and tolerance of abrasive solids, both products of an internal, moving seal.

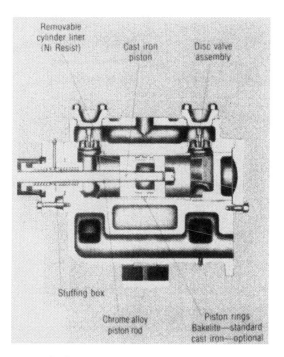

FIG. 8.27.   Piston pump liquid end. (Courtesy Worthington Pump, Dresser Industries Inc.)

Current practice for piston pumps tends to single acting, triplex power pumps and double acting, duplex direct acting pumps. These arrangements derive from the mechanics of the drive and a desire to reduce flow pulsations.

Piston pump speeds are governed by the following:

Power end; see Sections 8.17 and 8.18.
NPSHA
Viscosity
Piston and rod seal life; see Chapter 11.

The basic components of a piston pump liquid end are liquid cylinder, piston, valves, piston rod, and rod seal; see Fig. 8.27.

(a)

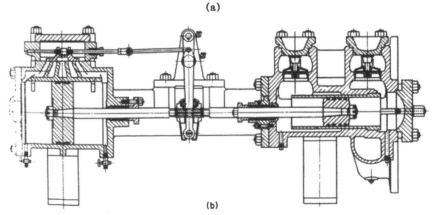

(b)

FIG. 8.28.   Side-pot type piston pump. (Courtesy Worthington Pump, Dresser Industries, Inc.)

Liquid ends are available in a variety of configurations. Those likely to be used in chemical processing are:

"Side pot," Fig. 8.28, are general purpose liquid ends able to work up to the pressure limit of piston pumps. Their suction and discharge valves have individual covers.

"Close clearance" liquid ends, Fig. 8.29, are used for volatile and gas-laden liquids. By minimizing the clearance at stroke end, the pump is able to handle liquids that would lower the capacity or vapor lock a conventional liquid end. Suction valves are located below the cylinder at the high point in the manifold to insure gas passes into the cylinder.

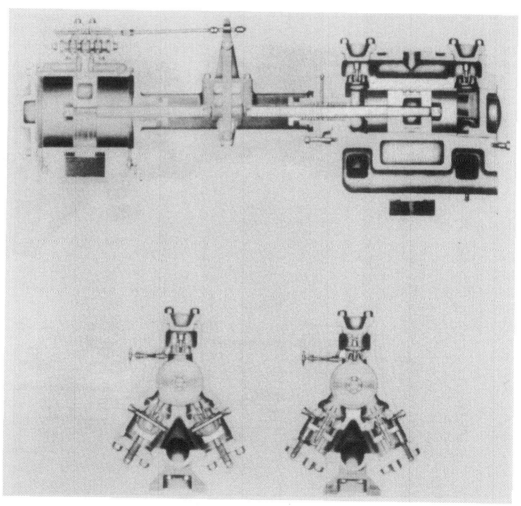

FIG. 8.29.   Close-clearance piston pump liquid ends. (Worthington Pump, Dresser Industries, Inc.)

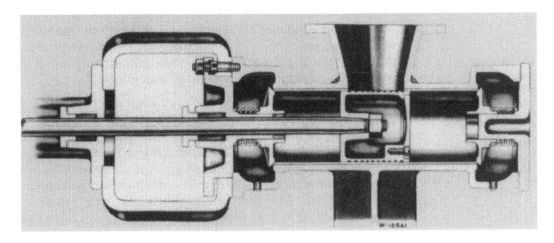

FIG. 8.30. Piston pump liquid end for high viscosity liquids. (Worthington Pump, Dresser Industries, Inc.)

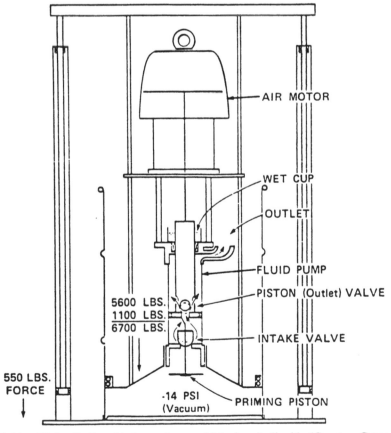

FIG. 8.31. Hydraulic forced inductor for very high viscosity liquids. (Courtesy Graco Inc.)

"High viscosity" liquid ends, Fig. 8.30, use the piston passing across a suction
port to eliminate suction valves, thus aiding flow into the pump. An
alternative approach is a direct acting pump with the pump inlet
immersed in the liquid: Vork [8.1] reports pumping liquids of 1.2 million
cP with the most sophisticated version of these, Fig. 8.31.

Liquid cylinders and covers are routinely produced in iron, bronze, and
cast steel. For special services, Ni-resist, cast nickel steel, chrome steels, and
chrome nickel steels are available. Cylinders are lined to allow restoration of
the bore. Liners are either thin wall, as in Fig. 8.30, or heavy wall, as in Fig.
8.28. Thin wall liners must be split to remove them, heavy wall liners can be
removed whole. Usual liner materials are iron, bronze, and Ni-resist.

Pistons can be split or solid, depending upon how the seal is mounted and
retained and the need to reduce weight. Four general seal types are employed:

Packing, Fig. 8.32, is used for water and similar liquids. The split piston
enables installation.
Metal rings, Fig. 8.33, angle cut, are used for oil and similar hydrocarbons.
Pistons can be solid or split.
Elastomer cups, Fig. 8.34, tend to wipe the bore clean. They are used when
the liquid contains abrasive solids. Pistons can be split or solid.
Serrations in the piston, Fig. 8.30, provide an adequate seal in high viscosity
liquids. For the pump shown in Fig. 8.30, serrations also avoid the risk of
seal damage in passing over the port.

Piston pump valves are self-acting. Four basic designs are available. The
choice of a particular design depends upon pressure, liquid viscosity, and
whether the liquid contains solids.

Stem guided disc. Fig. 8.35: As built for piston pumps, the disc valve is limited
to 350 lb/in.^{2}gauge. It is used for clean liquids to that pressure.

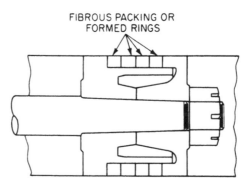

FIG. 8.32. Fibrous packing or formed ring piston seals. (Courtesy Worthington Pump,
Dresser Industries, Inc.)

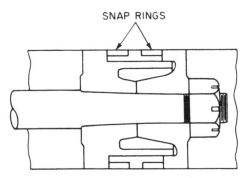

FIG. 8.33. Metal snap ring piston seals. (Courtesy Worthington Pump, Dresser Industries, Inc.)

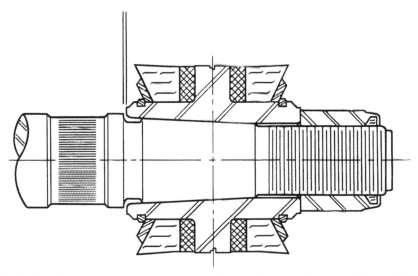

FIG. 8.34. Elastomer cup type piston seals. (Reprinted by permission of Mission Drilling Products Division of TRW Inc.)

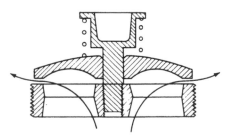

FIG. 8.35. Stem guided disc valve. (Courtesy Worthington Pump, Dresser Industries, Inc.)

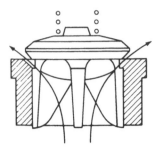

FIG. 8.36.   Wing guided valve. (Courtesy Worthington Pump, Dresser Industries, Inc.)

Wing guided valve, Fig. 8.36: Named for the wing guides which aid seating. The conical seat improves pressure tightness and reduces the risk of foreign particle obstruction. Wing guided valves are used for clean, low viscosity liquids to the pressure limit of piston pumps.

Ball or semihemispherical valve, Fig. 8.37: When open, these valves present a large area and a smooth passage, which aids the flow of viscous liquids. The valve's rolling action during closing affords a high tolerance of abrasive solids. Ball valves are usually free floating with a stop to limit opening. Semihemispherical valves are spring loaded. Both forms are good for the pressure limit of piston pumps.

Elastomer insert valve, Fig. 8.38: An alternative to the ball valve for abrasive service. Instead of aiming to clear the seat during closure, the elastomer insert valve seeks to accommodate solids by having one surface resilient.

Valve materials depend upon the valve type and its service conditions. Common materials are bronze, chrome steel, and chrome nickel steel. Severe services may warrant hard coating (chrome or cobalt alloy), tungsten carbide, and various nonmetallics,

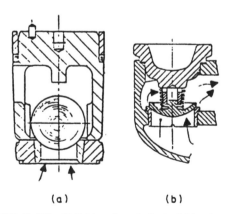

(a)   (b)

FIG. 8.37.   Ball (a) and semispherical (b) valves.

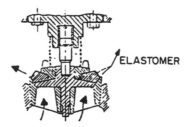

FIG. 8.38. Elastomer insert valve.

Piston rods are subject to compression, thus must be stiff enough to avoid buckling. Cyclic loading requires that fatigue be respected in their design and manufacture. The region under the rod seal has to be well finished and durable, and it is often hard coated to realize this.

Rod seals are discussed in Chapter 11.

For high capacity slurry service within the pressure capability of piston pumps, a surge leg arrangement can be employed. Figure 8.39 shows the principle. Pumped liquid, the slurry, is "separated" from the pump by a conduit filled with the slurrying liquor. Pump action on the liquor produces a similar action in the slurry, first ingesting it, then discharging it. Make-up liquor has to be injected into the separating conduit. Some liquor contamination should be allowed for in the pump construction.

A notable though still experimental piston pump development is a magnetically coupled, hermetically sealed design reported by Yeaple [8.2].

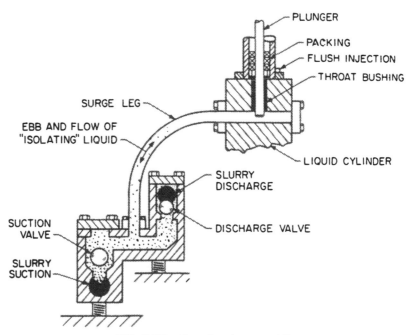

FIG. 8.39. Surge leg plunger pump.

8.20. Plunger Pumps

Plungers differ from pistons by having an external, stationary seal, see Fig. 8.40. This limits plungers to being single acting, but confers far greater pressure and solids handling capability.

Most plunger pumps are multiplex power pumps. Direct acting designs are available, usually simplex, but are now rarely used.

Plunger pump speeds are limited by a succession of factors:

Power end, see Fig. 8.17
Frequency of flow pulsation
NPSHA
Liquid viscosity
Plunger seal life related to surface speed and pressure drop; see Chapter 11

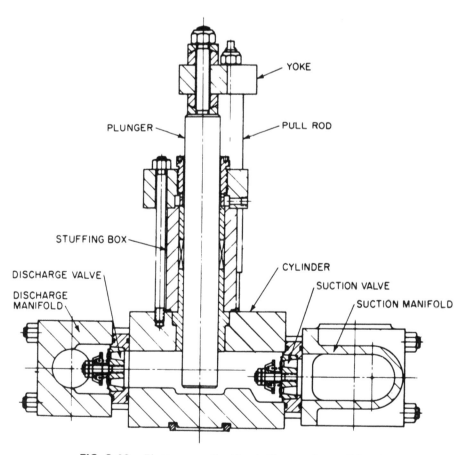

FIG. 8.40.   Plunger pump liquid end. (Courtesy Ingersoll-Rand.)

Buse, in Ref. 1.1, notes that plunger speed sometimes has to be limited to 140 to 150 ft/min for acceptable plunger seal life. Davidson [1.5] observes that plunger seal life governs pump availability. Davidson adds that pump cost increases with decreasing speed, thus producing substantial commercial pressure to use higher speeds.

A plunger pump liquid end consists of the following basic parts: cylinder block, plunger, valves and stuffing box; see Fig. 8.40. Three configurations of these parts are generally available:

"Standard," Fig. 8.41. A large clearance volume design with emphasis on good valve characteristics. Used for general service and liquids of high bulk modulus.

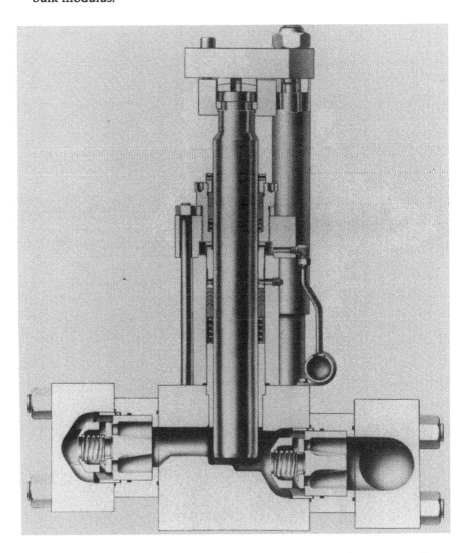

FIG. 8.41. "Standard" plunger pump liquid end. (Courtesy Worthington Pump, Dresser Industries, Inc.)

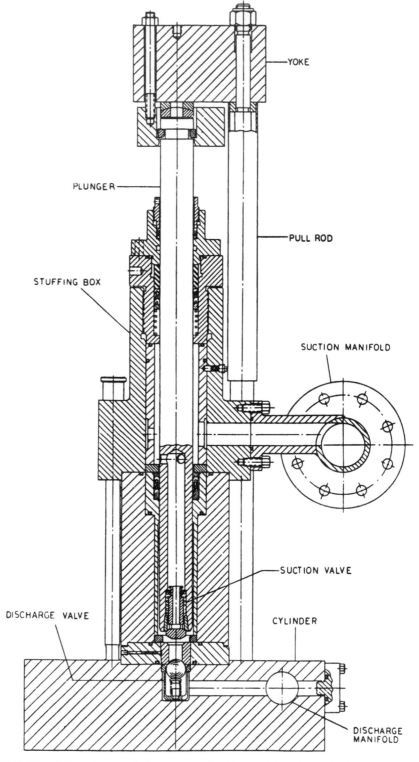

FIG. 8.42. "Close clearance" plunger pump liquid end with suction valve in plunger. (Courtesy Worthington Pump, Dresser Industries, Inc.)

"Close clearance," Fig. 8.42. By refining the valve arrangement, the clearance volume is reduced. The refinement is necessary when compressibility becomes a major factor in volumetric efficiency.

"Slurry," Fig. 8.43. A design for high solids concentration slurry. Parts are shaped for smooth flow. Valve type and arrangement are resistant to misoperation. Synchronized injection washes the plunger during each suction stroke.

Cylinder blocks are either one piece or individual. The virtue of a one piece cylinder block, Fig. 8.44, is integral suction and discharge manifolds. With individual valve covers, this enables valve maintenance without having to break the suction or discharge piping connections. Individual cylinder blocks, Fig. 8.45, are necessary when the pumping temperature exceeds 300°F in multiplex pumps. They are warranted for corrosive or high pressure service where manufacture to fine internal detail in exotic materials is required. Separate suction and discharge manifolds, special for high temperature, are required with individual cylinder blocks. In some designs these have

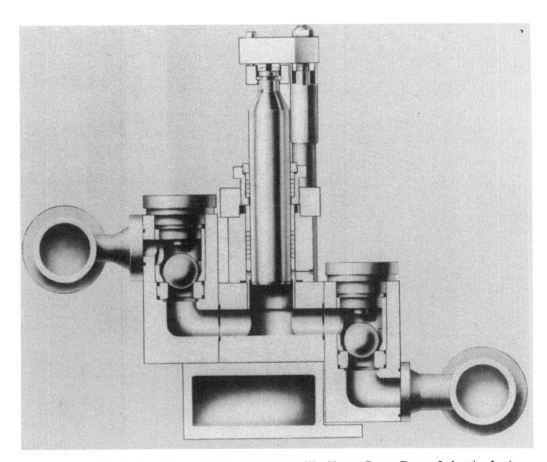

FIG. 8.43. "Slurry" liquid end. (Courtesy Worthington Pump, Dresser Industries, Inc.)

to be removed to service the valves. Other designs, Fig. 8.45, employ an arrangement allowing individual valve removal.

Low pressure cylinder blocks, e.g., up to 3000 lb/in.^{2}gauge in carbon steel, can be cast. Higher pressures require the integrity of forgings. Materials range from carbon steel through alloy steels to high alloys and aluminum bronze.

Plungers are solid to 5 in diameter, and hollow above that to reduce weight (Buse in Ref. 1.1). During the discharge stroke the plunger is loaded in compression. Small diameter plungers at high pressure should be checked for buckling. In vertical pumps the connection between plunger and yoke should have provision for centering the plunger in the stuffing box. Note, however, that this provision cannot compensate for angularity between the crosshead way and stuffing box bore. Plunger materials are hardened chrome steel, hard chrome, Colmonoy and ceramic on carbon steel, or 316 stainless steel and solid ceramic. Nonfused coatings are permeable, which can lead to spalling caused by either subcoating pressure or substrate corrosion.

Valves for plunger pumps are similar to those in piston pumps, see Section 8.19, but designed for higher pressures. Typical pressures are shown in Table

FIG. 8.44. Plunger pump with "one piece" cylinder block. (Courtesy Worthington Pump, Dresser Industries Inc.)

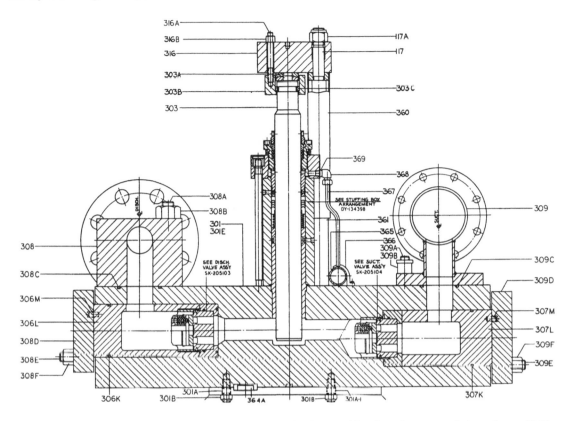

FIG. 8.45. Individual cylinder block with provision for valve removal through manifolds. (Courtesy Worthington Pump, Dresser Industries, Inc.)

8.5. Various designs incorporate refinements to improve valve guidance and sealing, see Fig. 8.46.

The stuffing box houses the plunger seal, see Chapter 11, and by way of the seal and throat bushing, serves to guide the plunger. Stuffing boxes are separate; integral designs would complicate cylinder block manufacture.

As with piston pumps, see Section 8.19, plunger pumps can be arranged with a surge leg for pumping slurry. The need to do so, however, is less given the plunger pump's slurry handling capabilities.

TABLE 8.5 Typical Reciprocating Pump Pressures[a]

Valve Type	Pressure (lb/in.^{2}gauge)
Plate (disc)	5,000
Skirt (wing)	10,000
Ball	30,000
Elastomer insert	2,500

[a]Source: Ingersoll-Rand; Buse in [1.1].

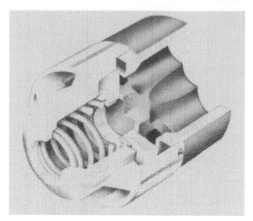

(a) outside guided plate valve

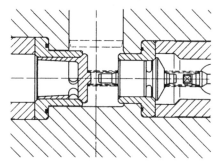

(b) outside guided skirt valve

FIG. 8.46. Outside guided plate and skirt valves. (Courtesy Worthington Pump, Dresser Industries, Inc.)

8.21. Diaphragm Pumps

The prime virtue of diaphragm pumps is the absence of a seal; the pumped liquid is contained and displaced by a flexible diaphragm. A secondary advantage is the ability to run dry. Figure 8.19 shows that there are two distinct classes of diaphragm pump: direct acting and power.

Direct acting pumps have the motive fluid, usually compressed air, applied to the drive side of the diaphragm. The pumps are duplex, see Fig. 8.47.

With compressed air as the motive fluid, the pressure and power of direct acting diaphragm pumps is limited. Despite this limitation, their sealless feature means wide usage for handling corrosive, toxic, or viscous liquids.

Figure 8.47 shows a typical direct acting diaphragm pump. Its major liquid end components are a pump casing, diaphragms, and suction and discharge valves.

Pump casings are made of metals and nonmetals. Metals include iron, aluminum, 316 stainless steel, Alloy 20, and Hastealloy 20. As with centrifugal pumps, nonmetals are replacing exotic alloys for corrosive service. Early pumps were constructed in GRP (graphite reinforced polymer). Nystrom and Larkin [8.3] report the use of polyvinylidene fluoride (PVDF) for pumps handling liquid chlorine and sulfuric acid.

The usual construction material for diaphragms is fabric reinforced synthetic rubber. Materials include neoprene, Buna N, butyl, and Viton. Teflon is used either as an overlay on a conventional diaphragm or as the diaphragm itself.

Three forms of valves are used: poppet, ball, and flap, in order of increasing tolerance of viscosity and solids size, see Fig. 8.48. Valves are elastomer or elastomer faced.

Power diaphragm pumps employ a power pump acting on a hydraulic system to actuate the diaphragm. Figure 8.49 illustrates the principle. These designs are intended to realize the virtues of power pumps, piston or plunger, while eliminating the seal. This they do, but at higher initial cost and at some expense to volumetric efficiency, the effect of hydraulic liquid compressibility.

While a diaphragm pump is hermetically sealed, there is the problem of leakage if the continually flexing diaphragm ruptures. The problem is addressed in two steps. First, design has been refined to yield diaphragms whose service life is predictable, thus allowing scheduled replacement prior to rupture. Dalley [8.4] cites the following examples:

Material	Pressure (lb/in.² gauge)	Temperature (°F)	Life (h)
PTFE	5,000	300	20,000
Metal	10,000	390	8,000

Vetter and Hering, in Ref. 1.2, note that PTFE diaphragms are tolerant of scratches (notches) but metal diaphragms are not.

When incidental leakage cannot be tolerated, a second diaphragm is introduced. Two arrangements are used: double diaphragm and "sandwich" diaphragm. Both have liquid between them, but in the latter the liquid is at atmospheric pressure unless one of the diaphragms ruptures. Should that happen, a pressure switch senses the pressure rise.

The usual power diaphragm pump arrangement is shown in Fig. 8.49; the liquid end is connected directly to the diaphragm housing. For pumping

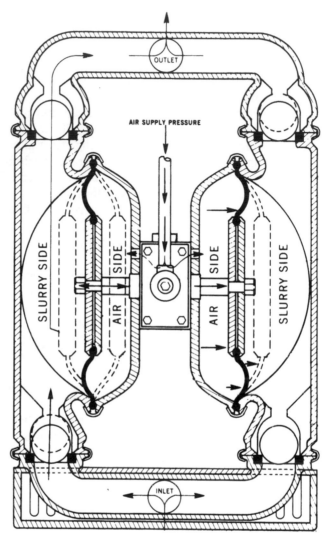

FIG. 8.47. Duplex diaphragm pump: major components. (Courtesy Wilden Pump & Engineering Co.)

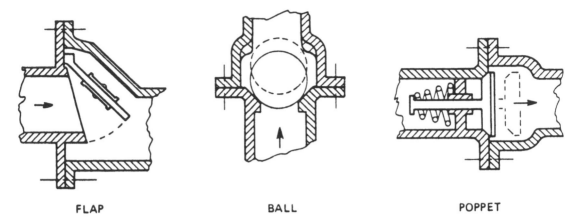

FLAP   BALL   POPPET

FIG. 8.48. Valve types for diaphragm pumps.

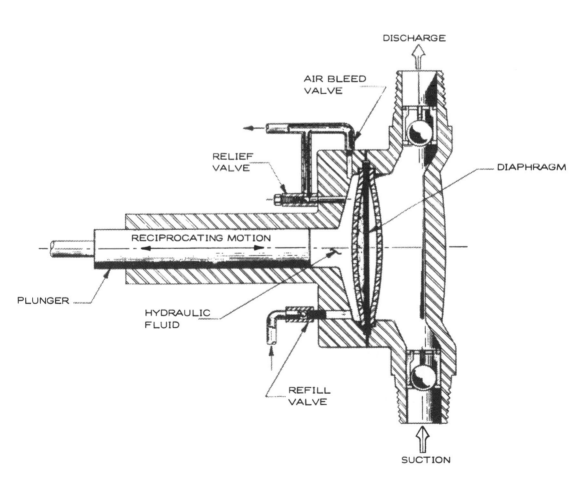

FIG. 8.49. Power diaphragm pump liquid end. (Courtesy Milton Roy.)

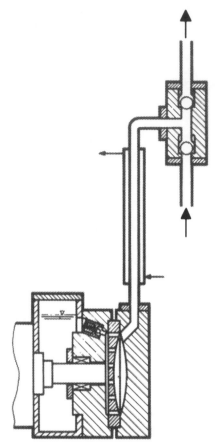

FIG. 8.50. Power diaphragm pump with liquid end located above diaphragm for high temperature service. (Courtesy LEWA.)

temperatures beyond the diaphragm's capability, the liquid end is located above the diaphragm housing, thus allowing a cool leg of pumped liquid to develop, see Fig. 8.50. When necessary, e.g., radioactive service, the entire "head" assembly (diaphragm and liquid end) can be located remotely, see Fig. 8.51.

Diaphragms are not restricted to the flat form shown in Fig. 8.49. Pearse [8.5] reports tubular diaphragm pumps, Fig. 8.52, operating successfully in slurry service.

To minimize flow pulsations, power diaphragm pumps are generally triplex or higher. Pearse [8.5], however, reports low speed duplex pumps with capacities up to 3300 gal/min.

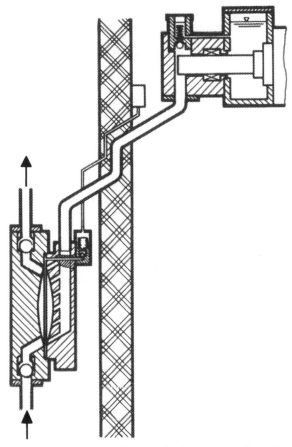

FIG. 8.51. Remote head configuration of power diaphragm pump for radioactive or similar liquids. (Courtesy LEWA.)

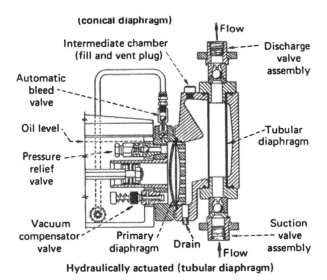

(conical diaphragm)

Intermediate chamber
(fill and vent plug)

Flow

Discharge
valve
assembly

Automatic
bleed
valve

Oil level

Tubular
diaphragm

Pressure
relief
valve

Vacuum
compensator
valve

Primary
diaphragm

Drain

Suction
valve
assembly

Flow

Hydraulically actuated (tubular diaphragm)

FIG. 8.52. Tubular diaphragm pump. (Courtesy Pulsafeeder.)

179

8.22. Mobile Separator

The arrangement termed "mobile separator" is a relatively complex pumping system which has been employed for high capacity slurry pumping at medium pressure (of the order of 600 lb/in.^{2}gauge). Figure 8.53 shows the simple flow diagram for one cylinder of what is usually a two or three cylinder system.

The essential feature of the system is separating slurry and drive liquor with a mobile separator. Slurry is charged into the cylinder by the feed pump, then discharged by drive liquor acting on the separator. Slurry and liquor flows are directed by valves whose operation is controlled by proximity switches at the cylinder ends. By using two or more cylinders, an essentially continuous flow can be produced. Because the drive liquor is clean, it can be handled by a multistage centrifugal pump.

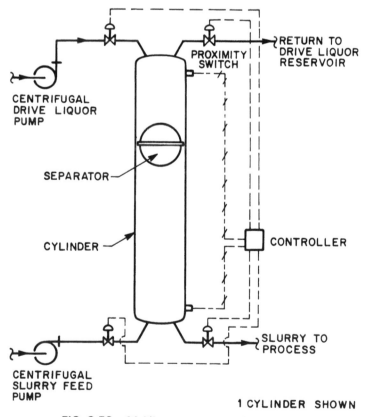

FIG. 8.53. Mobile separator-type reciprocating pump.

8.23. Industry Standards

API-674, Positive Displacement Pumps—Reciprocating [8.6] provides a working specification for some of the reciprocating pumps discussed in this chapter.

References

8.1. W. D. Vork, "Reciprocating Pumps for Viscous Materials," *Plant Eng.*, *31*(19) (September 15, 1977).
8.2. F. Yeaple, "Solar Energy, Hydrogen Sponge, Keys to Water Pump Operation," *Des. News*, pp. 255–257 (November 23, 1987).
8.3. M. Nystrom, and M. C. Larkin, "Fluropolymer Pumps Resist Corrosion," *Plant Serv.*, pp. 26–27 (September 1987).
8.4. G. Dalley, "Large Process Diaphragm Pumps, *World Pumps*, pp. 156–157 (April 1982).
8.5. G. Pearse, "Pumps for the Mineral Industry," *Min. Mag.*, *152*, 299 (April 1985).
8.6. *API-674, Positive Displacement Pumps—Reciprocating*, American Petroleum Institute, Washington, D.C.

9. Metering Pumps

Metering is the continuous or batch injection of controlled volumes of reactants into process streams. Pumps used for metering are either special versions of reciprocating pumps or conventional pumps with special control.

9.1. Pump Selection

Pump requirements are determined by:

Hydraulic duty, i.e., flow, differential pressure and reactant SG.
Necessary metering precision, related to allowable process variations, and the method of injection control.
Nature of the reactant, i.e., cleanliness, presence of large solids, viscosity, need to isolate from the atmosphere, and tolerance of dilution.

Figure 9.1 shows a broad approach to metering pump selection.

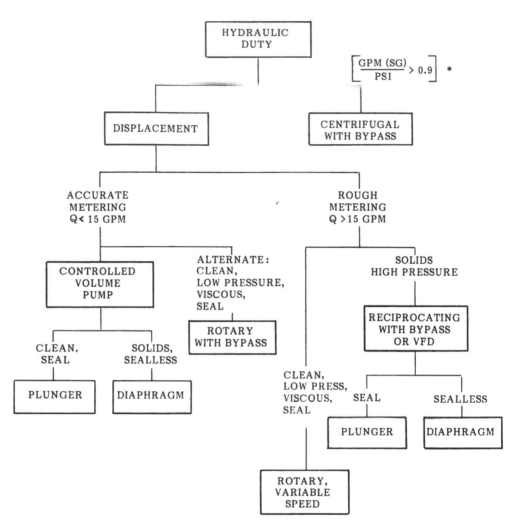

$$\left[\frac{GPM\ (SG)}{PSI} > 0.9\right]\ *$$

* FROM DAVIDSON, IN [1.5]

FIG. 9.1.   Metering pump selection.

Differentiating between displacement and centrifugal pumps can be done using either the chart cited in Chapter 5 or an empirical equation proposed by Davidson [1.5], see Fig. 9.1.

Within the hydraulic coverage of displacement pumps, accurate flow metering is defined as better than ±10%.

For accurate flow metering there are two approaches to pumping. The traditional approach is a variable displacement (controlled volume), reciprocating pump, see Fig. 9.4. Trent, in Ref. 1.2, advocates a fixed speed rotary pump with bypass control for services involving clean, viscous reactants at injection pressures up to 100 lb/in.^{2}gauge. The latter approach allows a less expensive pump but requires more system (bypass valve, piping, and a large suction vessel). A bypass continually agitates the reactant, which can be an advantage.

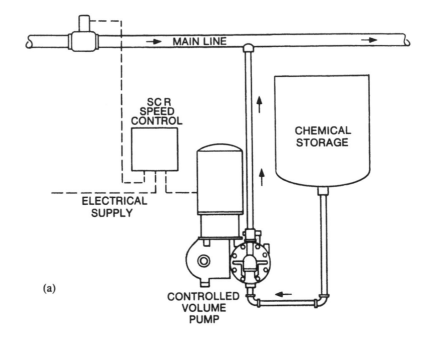

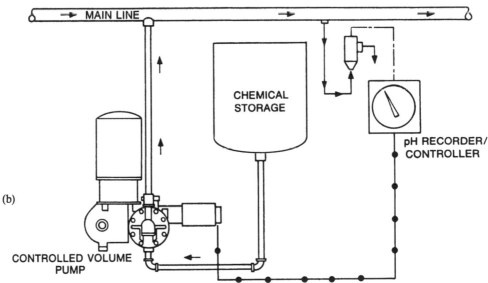

FIG. 9.2. Open (a) and closed (b) loop control in which the metering pump serves as the final control element. (Courtesy Milton Roy Co.)

"Rough" metering is done with a variable speed rotary pump whenever the reactant characteristics and injection pressure permit.

Reciprocating pumps, for both "accurate" and "rough" metering, are of the plunger type when seal leakage or reactant dilution is allowable. When not, diaphragm pumps are required.

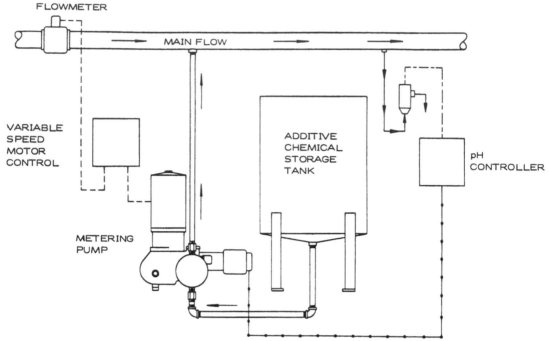

FIG. 9.3.   Combined open and closed loop control for wide swings in process flow. (Courtesy Milton Roy Co.)

9.2. Control

Metering control can be open loop, closed loop, or combined. Open loop control, Fig. 9.2(a), injects reactant in proportion to process flow. Poynton, in Ref. 1.2, notes that in open loop control, reactant flow does not have to go to zero, thus flow control can be by speed variation. Closed loop control, Fig. 9.2(b), measures the effect of reactant injection and modulates injection accordingly. Because the injection flow can go to zero, closed loop control varies pump displacement by stroke length or bypass opening. When process flows can vary widely, controlled volume pumps require combined open and closed loop control to produce the required reactant flow range, see Fig. 9.3.

9.3. Controlled Volume Pumps

Liquid ends are either plunger or diaphragm; see Chapter 8. Power ends can be slider crank, cam and follower, or direct acting. Variable stroke operation of the first two configurations is achieved one of two ways:

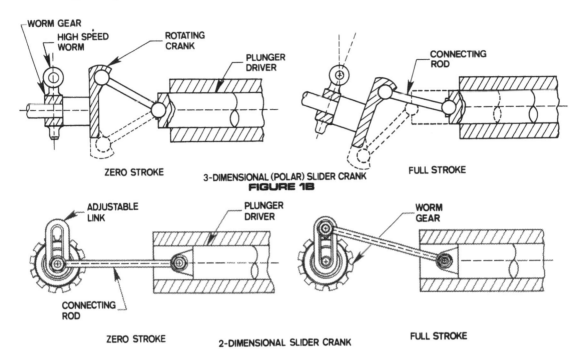

FIG. 9.4. Variable displacement (controlled volume) reciprocating pumps employing adjustable slider crank mechanisms. (Courtesy Milton Roy Co.)

Adjustable slider crank: a mechanism, either two or three dimensional, see Fig. 9.4, allows adjustment of the effective crank length.

Lost motion; mechanical: in cam and follower power ends, an adjustable stop limits the plunger's return stroke, see Fig. 9.5(a).

Lost motion; hydraulic: in hydraulically actuated diaphragm pumps, part of the power end displacement is bypassed through an adjustable port, see Fig. 9.5(b).

Typical capabilities of controlled volume pumps are tabulated in Table 9.1

Accuracy is that attainable on a test stand. Davidson [1.5] suggests in service accuracy close to ±2.5%, repeatability ±2.0%, all with clean gas-free liquid, adequate NPSHA, and stable service conditions. For good accuracy over 0 to 20% of rated capacity, both Davidson and Poynton note the need to vary stroke length and frequency.

Most controlled volume pumps are simplex, thus the flow pulsates significantly. System corrections and pulsation dampening, when necessary, are made in the same manner as those for larger reciprocating pumps; see Section 8.14.

Drivers are generally squirrel cage electric motors. When speed variation is required, a variable frequency control is introduced between the motor and power supply.

Direct acting drives, using compressed air, are employed when suited to the process or control, or when the environment is hazardous.

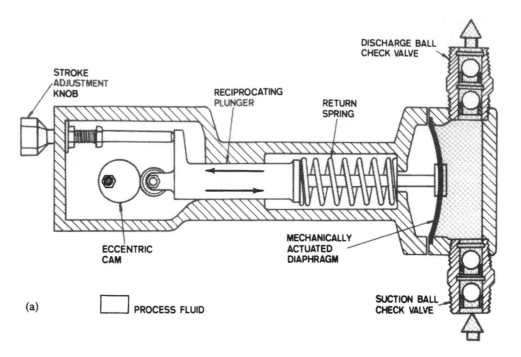

(a)   PROCESS FLUID

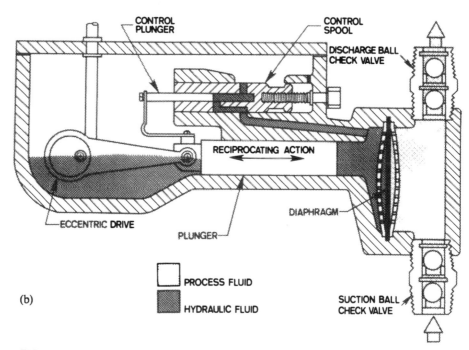

(b)   PROCESS FLUID

       HYDRAULIC FLUID

FIG. 9.5. Two types of lost motion drives: eccentric cam (a) and hydraulic bypass (b). (Courtesy Milton Roy Co.)

TABLE 9.1 Controlled Volume Pump Capabilities[a]

Type	Flow gal/min	Accuracy/Range	Pressure (lb/in.²gauge)
Plunger	20	1% over 15:1	50,000
Diaphragm—mechanical[b]	0.25	≃5%	150
Diaphragm—hydraulic[c]	20	1% over 10:1	5,000

[a]Source: Glanville in Ref. 1.1.
[b]10% zero shift over working pressure range.
[c]3 to 5% zero shift per 1000 lb/in.². Vetter and Hering, in Ref. 1.2, cite micrometering designs capable of 10,000 lb/in.²gauge.

9.4. Rotary Pumps

Trent, in Ref. 1.2, notes the use of gear pumps for metering. Poynton, also in Ref. 1.2, includes screw, vane, lobe, and peristaltic pumps. Chapter 7 covers rotary pumps application and selection.

9.5. Reciprocating Pumps

Conventional plunger or hydraulically actuated diaphragm pumps are used for "rough" metering. Chapter 8 details reciprocating pumps application and selection.

9.6. Centrifugal Pumps

Most metering services involve hydraulic duties outside centrifugal pump coverage. When the duty is suitable, see Fig. 9.1, refer to Chapter 6 for guidance on application and selection.

Reference

9.1. *API-675, Reciprocating Pumps—Controlled Volume*, American Petroleum Institute, Washington, D.C.

10. Utility Pumps

In a chemical processing plant, utility services cover the generation of electricity and process steam and the provision of coolant for steam condensing and other process cooling. The pump services involved are fuel, condensate, boiler feed, and cooling water circulation. Figure 10.1 shows a system with all the likely elements: a fired boiler, combined pass-out and condensing turbine, open cycle liquid side, and cooling tower heat dissipation. Other arrangements, e.g., waste heat boiler and backpressure turbine, have less liquid handling components.

10.1. Fuel Pump

Pumping fuel oil to medium and large boilers is almost exclusively the province of triple screw pumps. High volumetric efficiency, tolerance of a wide range of oil viscosities, and some tolerance of dirt account for this. Further details on the pump type are given in Chapter 7.

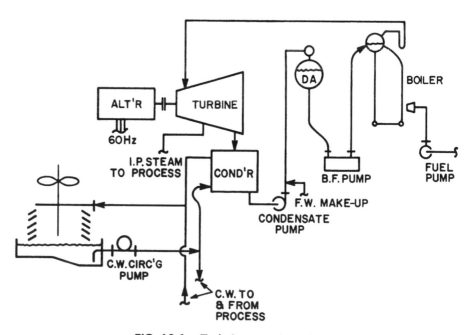

FIG. 10.1. Typical process plant utilities.

10.2. Condensate Pump

When the steam turbine is condensing, a pump is necessary to pump the condensate from the hotwell in the first stage of raising the liquid to boiler pressure. In an open cycle, as in Fig. 10.1, the condensate pump discharges into the deaerator. The alternative, a closed cycle, does not have a separate deaerator, so the condensate pump discharges directly into the feed pump suction.

Taking suction from the hotwell, the NPSHA at the condensate pump is the submergence less the piping friction loss. To avoid deep basements, condensate pump submergence is kept to a minimum, thus the NPSHA is low. Two approaches are used to accommodate this circumstance. The first, and older, approach is a dry pit pump running at low speed. To develop the required total head, the low speed pump has to be either multistage or operate in series with a separate condensate booster. Separate pumps complicate the drive but avoid the use of a special, and therefore expensive, multistage condensate pump. The suction of dry pit condensate pumps must be vented to the hotwell vapor space; otherwise the pump may become vapor bound. Figure 10.2 shows a typical horizontal condensate pump.

The second approach is to use a vertical can pump to increase the NPSHA at the first stage impeller. With more NPSHA, the pump can run faster and therefore be smaller. Vertical can pumps are used in other chemical processing applications where NPSHA at grade is low; see Chapter 6. Service life is

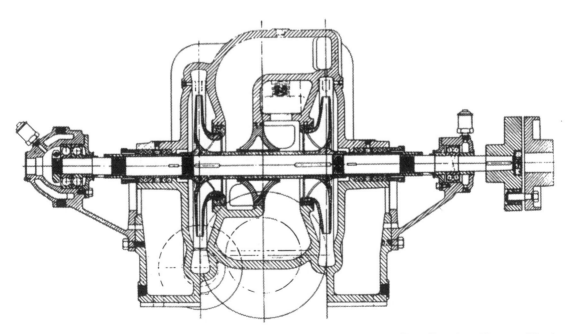

FIG. 10.2.   Two-stage horizontal condensate pump with axilly split casing. (Courtesy Worthington Pump, Dresser Industries, Inc.)

generally shorter than that of equivalent dry pit pumps, but capital costs, civil and equipment, tend to be lower.

Condensate pumps typically have relatively high S first stage impellers, and thus need to have high minimum flows if noise, vibration, and premature impeller erosion are to be avoided. See Chapter 6 for more details.

A refinement of the first approach, applicable to flows up to 120 gal/min, is a 3550 r/min single stage, overhung pump with an inducer. The higher rotative speed enables the head to be developed in a single stage. Because the energy is low, the pump can be operated down to 40 to 50% of BEP. Lower cost is the objective of this refinement.

Condensate pumps have iron casings. Impellers, wearing rings, bushings, and sleeves are either bronze or 13 chrome, the choice depending upon exact condensate chemistry. Where 316 stainless steel impellers are available as standard, they can be used instead. Shafts can be carbon steel or 13 chrome. Steel shafts should be fully covered within the pump.

Shaft seals in condensate pumps are usually arranged to be at a positive pressure, 15 $lb/in.^2$gauge minimum, when the pump is running. With this arrangement there will be no air in leakage during normal operation. Some provision is necessary, however, to ensure no air in leakage when a shutdown pump is exposed to condenser vacuum.

10.3. Boiler Feed Pump

The boiler feed pump has to develop sufficient pressure to put water into the boiler drum after overcoming friction losses in the interconnecting piping and system. With centrifugal boiler feed pumps it is important to recognize that the pressure is developed with water at the pumping temperature, not the drum temperature.

Boiler feed pump NPSHA is provided by deaerator elevation in an open cycle, as in Fig. 10.1, by the condensate pump in a closed cycle. During sudden load drops, the NPSHA in an open cycle can momentarily fall well below the nominal value (under the action of incoming cold condensate without commensurate heat input, deaerator pressure has fallen below the vapor pressure of hotter liquid still in the suction line), thus an NPSH margin of some 50% of NPSHR is typical.

Both displacement and kinetic pumps are used for boiler feed service, the choice being based on hydraulic duty, flexibility, and maintenance.

Displacement pumps are generally of the horizontal plunger type. Their use is restricted to small, high pressure boilers. Capacity control is achieved with a bypass or by speed variation. Chapter 8 includes information on plunger pumps.

Kinetic pumps can be regenerative turbine, multistage, or high speed single stage. Regenerative turbine pumps are rarely used; see Chapter 6.

Multistage centrifugal pumps offer the greatest versatility and are therefore widely used. Chapter 6 outlines the types available and their selection.

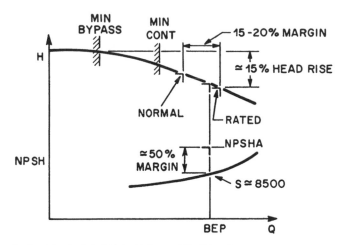

a) GOOD SPECIFICATION & SELECTION

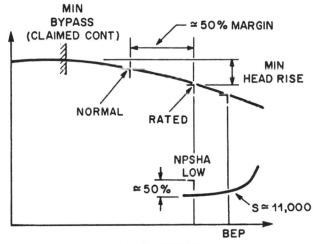

b) POOR SPECIFICATION & SELECTION

FIG. 10.3.   Boiler feed pump specification and selection.

There are, however, three points of special note on hydraulic specification and selection:

Rated capacity. The margin over normal flow should be kept small; 15% is tolerable but 50% is potentially troublesome since the pump will be oversized. Rated capacity should be at or beyond BEP, thus allowing normal operation close to BEP.

NPSHR. Of necessity, boiler feed pumps are high energy pumps and many have to run over a wide flow range as a plant load swings. To do this without distress, the pump must be sized correctly (see above) and its suction specific speed should not exceed 8500 to 9000. The NPSHR for

S = 8500 to 9000 plus the need for a substantial margin (see above) should be allowed for in preliminary plant design.

Control. Adequate control requires a well-defined intersection between the pump and system characteristics. To realize this, the pump characteristic must be sufficiently steep or the usually flat system characteristic artificially steepened by throttling. The latter wastes energy, so the pump is generally required to have at least 10% head rise from rated to minimum bypass flow. Conventional designs selected with rated flow at or beyond BEP readily meet this requirement. Note that the head rise is specified back to minimum bypass flow. The pump cannot run below this flow, so its head characteristic in that region is not important.

Figure 10.3 illustrates the points made above. Note that a "poor" selection is often less expensive (less stages) and more efficient at rated capacity (larger pump), and thus is superficially the better selection. When this artifice is not precluded by the specification, commercial pressure will dictate that all manufacturers at least offer the "poor" selection.

TABLE 10.1   Boiler Feed Pump Materials

Temp. F	pH	O$_2$ ppm	Conductivity microhms/cm	Materials (code below)[a]					
				I-1	I-2	S-1	S-6	C-6	A-8
≤200	0–14	—	—						☐
	4.5–14	—	—					☐	☐
	6– 9	—	—		☐[b]		☐	☐	☐
	6–14	—	—				☐	☐	☐
	9–14	—	—	☐		☐[b]	☐	☐	☐
>200	0–14	>0.04	≤20						☐
	4.5–14							☐	☐
>200	0–14	≤0.04	≤20						☐
	4.5–14							☐	☐
>200	0–14	≤0.04	>20						☐
	4.5–14							☐	☐
	6– 9				☐[b]		☐	☐	☐
	6–14						☐	☐	☐
	9–14			☐[b]		☐[b]			

[a]*Note*: Many chemistries marginal; avoid selecting less expensive materials when C-6 indicated.

Code (API-610)	Casing, Impeller
I-1	All cast iron.
I-2	Cast iron, bronze.
S-1	Cast steel, cast iron.
S-6	Cast steel, 13 chrome.
C-6	All 13 chrome.
A-8	All CF-3M.

[b]Head less than 400 feet/stage.

High speed single stage pumps are an alternative to multistage pumps for smaller boilers. Selection requires care to insure that the expected flow range is within the pump's capability. Partial emission pumps have flat head characteristics which need to be checked for controllability.

Materials for boiler feed pumps should be chosen with care. Feed water temperature and chemistry (pH, O_2 conductivity) are fundamental to making the correct choice, as shown in Table 10.1.

10.4. Cooling Water Pumps

The cooling water pump takes its suction from a natural water body, an artificial water body, or a cooling tower basin. The last is the most likely source in modern plants.

Cooling water circulation is a high flow, low head duty, thus exclusively a centrifugal pump service. Two pump configurations are used: dry pit with horizontal and vertical axes, and vertical wet pit. A typical horizontal dry pit pump is shown in Fig. 10.4 Wet pit pumps are better if of the volute type with an enclosed lineshaft, see Fig. 6.60. The virtue of this configuration is greater bearing integrity.

FIG. 10.4.   Installation of double suction, horizontal circulating pumps. (Courtesy Worthington Pump, Dresser Industries, Inc.)

TABLE 10.2 Cooling Water Pump Materials

			Materials	
Water	Environment	Casing	Impeller	Steel
Fresh	Normal	Iron	Bronze	Steel
Fresh	Basic (no Cu alloys)	Iron	13 Chrome	Steel
Salt, <80°F	Normal	Niresist	bronze	Monel
Salt, <80°F	Basic	Niresist	CF8M	Duplex

The rated capacity of cooling water pumps is frequently a victim of "margins upon margins," leading to gross oversizing. When the excess capacity is "throttled off," the result can be cavitation-like noise, vibration, thrust bearing failure (associated with rotor shuttling), clearance wear (in single volute pumps), and premature impeller erosion. Rated capacity should be determined by using only the margin necessary. In cases where the flow can vary widely, e.g., one pump supplying multiple units, provision should be made in the plant design to keep the gross flow within the pump's continuous operating range.

NPSHA is the other aspect of cooling water pump specifications that frequently suffers from "margins upon margins." The typical case understates NPSHA (see Chapter 6), then requires a further margin, thus forcing the selection of a high suction specific speed pump. Such a selection, of course, has even less tolerance of the oversizing identified above.

Pumping from aerated sources can lead to a drop in performance caused by air coming out of solution in the low pressure regions of the pump. As noted in Chapter 4, this is not cavitation but does require a correction to NPSHA if the performance drop is to be avoided. The inclusion of this correction should be noted in the pump specification.

The choice of materials depends upon the water and plant environment. Broad guidelines are given in Table 10.2. See Fraser, in Ref. 1.1, for additional information.

Except for very small pumps, iron impellers should not be used in cold water service. Iron has poor resistance to cavitation erosion, and in large pumps the extent of localized cavitation (with no discernible performance drop) is often sufficient to cause premature impeller failure.

CF8M (cast 316 stainless steel) is prone to pitting corrosion in stagnant seawater, which could present a problem in a standby pump. The alternative is to use a duplex alloy impeller.

Shafts for seawater pumps are Monel or duplex alloy in order to avoid the low strength limitations of 316 stainless steel. The effect of a corrosive environment on endurance strength (see Chapter 3) compounds the problem.

11. Seals

All pumps are sealed. Most have some form of dynamic liquid seal where the moving element passes through the pressure boundary. Such seals all leak to some degree, though often at very low rates, and the leakage usually goes to "atmosphere." When leakage is unacceptable or sealing is difficult, pumps designated "sealless" are often used. Actually, these pumps are either specially arranged to avoid a liquid seal or hermetically sealed.

Because the construction of "sealless" pumps is special, they're covered in the chapters dealing with each pump type (6, 7, and 8), with chapter 6 having a detailed description of the two basic arrangements. The various devices used for dynamic liquid seals are covered in this chapter. For completeness, selection charts also cite both forms of "sealless" pump where appropriate.

A pump's seal is a fundamental element of its construction. Because of the generally poor service life of seals in chemical processing, typically only 4000 to 5000 h, it is clear the seal needs to be treated as an integral part of the pump's design.

11.1. Environment

The greatest single influence on seal performance is its environment. A very ordinary seal in an appropriate environment will run whereas a sophisticated seal in the wrong environment will not.

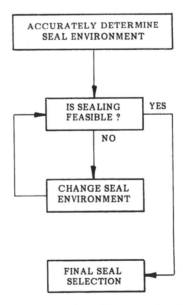

FIG. 11.1. Seal selection—fundamental consideration.

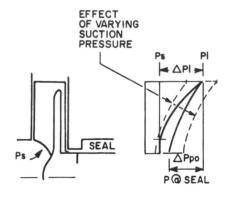

EFFECT
OF VARYING
SUCTION
PRESSURE

$P @ SEAL = Ps + \triangle Pi - \triangle Ppo$

$Ps = $ SUCTION PRESSURE

$\triangle Pi = $ IMP PRESS RISE

$\triangle Po = $ P.O. VANE PROSS DROP

$\triangle Po$ CONSTANT, $\triangle Pi$ VARIES

WITH Q , $\therefore$ P @ SEAL VARIES

FROM $<$ Ps TO $>$ Ps

(a)   SEMI - OPEN IMPELLER

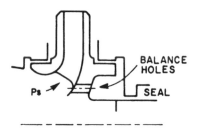

$P @ SEAL \simeq Ps$

(PROVIDED BALANCE HOLES

LARGE ENOUGH).

WHEN MULTISTAGE, $Ps = $ SUCTION

PRESSURE OF STAGE

(b)   SINGLE SUCTION, BALANCED IMPELLER, SEAL BEHIND

FIG. 11.2.   Pressure at the seal: (a) Semi-open impeller; (b) single suction, balanced impeller, seal behind; (c) single suction, unbalanced impeller, seal behind; (d) single suction, unbalanced impeller, seal ahead.

A seal's environment is described by the following:

Pressure
Temperature
Liquid condition
Surface speed
Displacements

Seal selection involves establishing the actual environment, checking seal feasibility, then iterating through environment changes if necessary. Figure 11.1 illustrates the concept.

Implicit in the question "Is sealing feasible?" in Fig. 11.1 is the qualification "to the required degree." In some circumstances the required degree may be subject to individual tolerance; in others, to legislation.

To be sure the seal's environment is accurately determined, it is worth reviewing each of its elements.

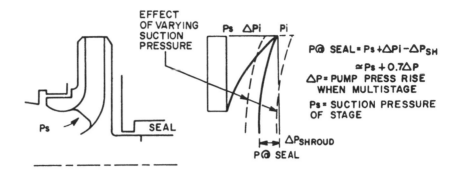

$$P @ \text{SEAL} = Ps + \Delta Pi - \Delta P_{SH}$$
$$\cong Ps + 0.7 \Delta P$$
$\Delta P =$ PUMP PRESS RISE
WHEN MULTISTAGE
$Ps =$ SUCTION PRESSURE
OF STAGE

(c)   SINGLE SUCTION, UNBALANCED IMPELLER, SEAL BEHIND

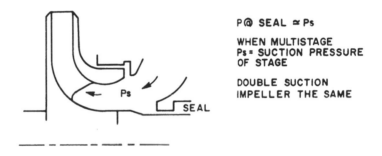

$$P @ \text{SEAL} \cong Ps$$

WHEN MULTISTAGE
$Ps =$ SUCTION PRESSURE
OF STAGE

DOUBLE SUCTION
IMPELLER THE SAME

(d)   SINGLE SUCTION, UNBALANCED IMPELLER, SEAL AHEAD

FIG. 11.2.   Continued

Pressure. The pressure at the seal during operation. Pump configuration and internal flow have a major influence on this pressure, so it should be checked carefully. Centrifugal pumps are the most difficult to analyze. Figures 11.2(a), (b), (c), and (d) show the usual configurations. When the pump is shut down, the pressure at the seal can be quite different and may dictate special precautions to avoid pump or process problems.

Temperature. The temperature at the seal during operation, including any necessary allowance for seal friction heat. Normally the temperature at the seal is essentially pumping temperature. In some circumstances, however, it can be higher, e.g., if the seal is adjacent to a major breakdown bushing in a high pressure multistage pump.

Liquid condition. Chapter 2 details the data needed to describe the liquid. Particular attention should be paid to suspended solids, viscosity, freezing point, tendency to crystallize, and chemical stability.

Surface speed. The speed at the effective sealing surface. In reciprocating pumps, the mean is used.

Displacements. During operation, the shaft or rod being sealed may be displaced. Some industry standards, e.g., ANSI B73.1 [6.21] and API-610 [6.24], limit the displacements. Others do not, and then it is important to establish the limits.

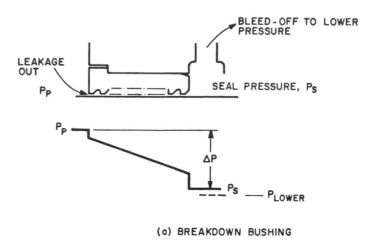

(a) BREAKDOWN BUSHING

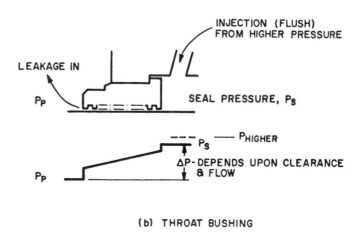

(b) THROAT BUSHING

FIG. 11.3. Means of varying pressure at seal: (a) breakdown bushing to lower pressure; (b) throat bushing to raise pressure.

Changes to the seal environment are usually made using the following techniques.

Pressure. Reduction: Flow through a breakdown bushing, Fig. 11.3(a), to a point of lower pressure lowers the pressure at the seal. The actual reduction depends upon bushing geometry and the size of the bleed-off. Bushing wear increases the pressure at the seal. Increase: Changing impeller configuration by deleting balancing holes (Fig. 11.2b to 11.2c) raises the pressure at the seal. When liquid conditions allow it, eliminating pump-out or clearing vanes (Fig. 11.2a) has the same effect. Both these modifications affect axial thrust, so they can only be made once the thrust bearing capacity has been verified. The alternative is to inject sufficient flow through a throat bushing (Fig. 11.3b). The flow necessary can be quite high, particularly when pumping low SG liquids, and its injection often requires special attention to avoid ill effects on the seal. Throat bushing wear lowers the pressure at the seal.

Temperature. Reduction: In most pump applications, the ideal is to have the seal at pumping temperature. Doing this avoids complicating the pump and its installation. Characteristics of the liquid or the seal or both, however, frequently require that the liquid temperature at the seal be lowered to realize adequate life or, in some cases, just to survive at all. Five

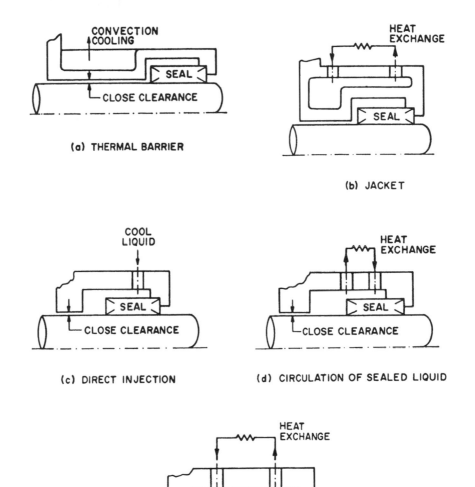

FIG. 11.4. Means of lowering temperature at seal: (a) thermal barrier; (b) cooling jacket; (c) direct injection; (d) circulation of sealed liquid through cooler; (e) circulation of intermediate liquid through cooler.

means of lowering seal temperature are in general use. They are thermal barrier, jacket, direct injection, circulation of the sealed liquid, and circulation of an intermediate or buffer liquid; see Figs. 11.4(a) through (e). Table 11.1 summarizes the advantages and disadvantages of each. Increase: When the liquid can crystallize, precipitate a solid, or become highly viscous as its temperature falls, the need can arise to increase the temperature of the liquid at the seal. Circulating a heat transfer fluid through a jacket around the seal is the usual means of doing this.

Liquid condition. Liquids containing abrasive solids or liquids subject to chemical change in passing through a seal are better kept out of the seal. In essence this is done by opposing the pumped liquid with a clean and stable liquid at higher pressure. Direct injection through a throat bushing (Figs. 11.4c and 11.3b) is one arrangement. To be effective, the liquid velocity through the throat bushing clearance has to be at least 7 ft/s but is better at 15 ft/s. A pressurized buffer liquid between low leakage seals is an alternative approach which offers lower dilution but complicates the seal. A second aspect of liquid condition is the effect on the atmospheric side of the seal. When there is a risk of scaling, carbonizing, crystallizing, corroding, or gasifying, a "quench" fluid is circulated around the atmospheric side of the seal to keep the seal clear or to dilute and remove leakage.

Surface speed. A function of pump size and speed, surface speed cannot be varied for a given pump selection, but it can serve to limit the pump selection or the type of seal.

Displacement. When not mandated by an industry standard, abnormal displacements of the element being sealed are a compromise between pump and seal design.

TABLE 11.1 Advantages and Disadvantages of Cooling Seals.

Cooling Means	Advantages	Disadvantages
Thermal barrier	No contact with pumped liquid No auxiliary services	Limited effectiveness Require extra axial length
Jacket	No contact with pumped liquid	Require auxiliary service Limited effectiveness Prone to scaling
Direct injection	Most effective cooling Low cost	High heat loss Coolant must be compatible with pumped liquid
Circulation of pumped liquid	Minimum heat loss Primary coolant is pumped liquid	Require secondary heat exchange Higher cost Risk of shaft bowing in hot standby pumps
Circulation of buffer liquid	Limited contact with pumped liquid Allows use of effective coolant	As "circulation of pumped liquid"

11.2. Seal Types and Selection

The seal types used in pumps for chemical processing and associated services are shown in Fig. 11.5. The motion of the element being sealed has a major bearing on seal form, construction, and operation, hence the distinction between rotating and reciprocating sealed elements.

Seal selection is a complex task. Selecting the basic type is the first step. Figure 11.6 categorizes seal type by allowable leakage, which is often a prime consideration. Figures 11.7 and 11.8 set down appropriate seal types for various liquids and service conditions. Once a basic seal selection is made, it is refined for the actual service conditions by using data for the type. Seal technology is continually evolving and its application is often subjective, so any selection process can only be for general guidance.

11.3 Soft Packing (Packed Box)—Rotating Shaft

The shaft to be sealed is surrounded by an annular region known generally as a "stuffing box"; see Fig. 11.9. Soft packing, usually in separate rings, is installed in the stuffing box, then compressed axially with a gland to produce a very close radial clearance between the shaft and packing. Packed box seals must have some leakage between the shaft and packing to lubricate the seal and to remove the small amount of heat generated.

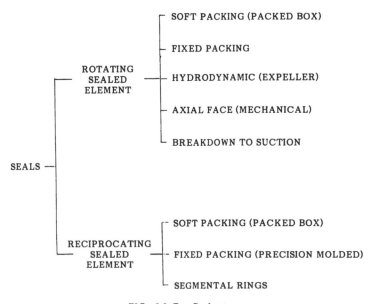

FIG. 11.5. Seal types.

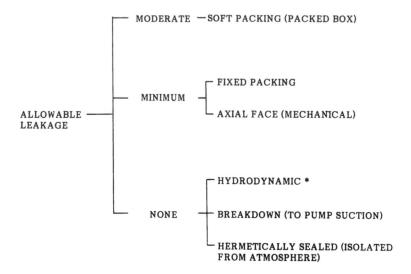

* ZERO LEAKAGE WHEN RUNNING

FIG. 11.6.   Seal type versus allowable leakage.

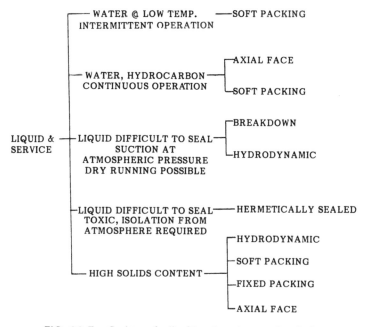

FIG. 11.7.   Seal type for liquid and service; rotating shaft.

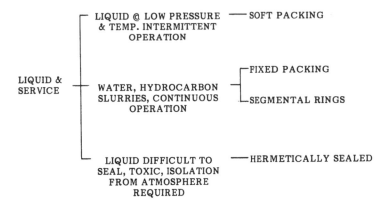

FIG. 11.8. Seal type for liquid and service; reciprocating.

The upper limit of packed box seal capability is given in terms of pressure versus surface speed in Fig. 11.10. Applications above these limits are prone to rapid packing and sleeve wear and high leakage. Breakdown bushings, Fig. 11.3(a), allow a packed box when the pressure adjacent to the seal exceeds those in Fig. 11.10.

Pressures at the seal below 5 lb/in.^{2} gauge pose another problem. Seal leakage will be little or none, resulting in overheating and, when the pressure is below atmospheric, air leakage into the pump. Installing a seal cage in the packing assembly, Fig. 11.11, and injecting sealing liquid overcomes the problem.

Solids-laden pumped liquids have to be excluded from the seal. Combining a throat bushing and a seal cage, and injecting clean liquid, Fig. 11.12, is the most effective solution, but at the expense of product dilution. If minimum product dilution is important, an arrangement similar to Fig. 11.11 can be used. Packing and sleeve wear will be higher, however.

High pressure adjacent to the seal plus solids in the pumped liquid requires a clean liquid-injected breakdown bushing, Fig. 11.13.

The pressure drop through a packed box seal is not linear. Because rings nearer the low pressure end carry a greater proportion of the total seal load,

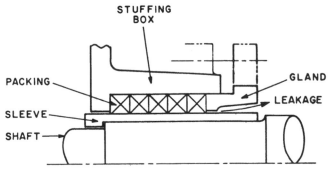

FIG. 11.9. Packed box seal.

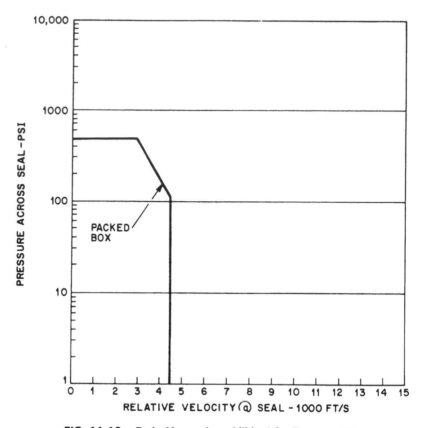

FIG. 11.10. Packed box seal capabilities (after Durametallic).

they are compressed more and run closer to the sleeve, hence have a higher pressure drop; see Fig. 11.14. The import of unequal pressure drops is that beyond, say, four or five rings, extra packing makes little contribution to seal capability and can in fact detract from it by generating more heat. Figure 11.14 shows the redundancy of three rings in an eight-ring seal. Flitney [11.1] notes that recent research confirms the feasibility of fewer packing rings while also revealing a tendency for the pressure drop to become closer to linear as the packing "wears in."

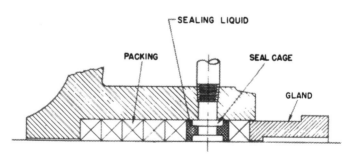

FIG. 11.11. Conventional packed box with seal cage.

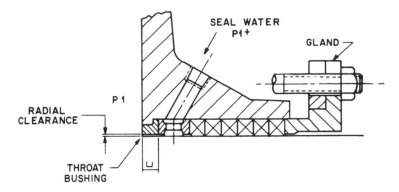

FIG. 11.12.   Packed box with throat bushing for clean liquid injection.

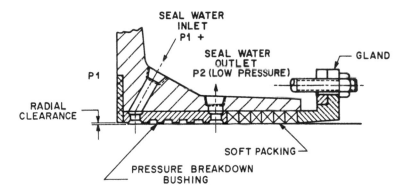

FIG. 11.13.   Injected breakdown bushing.

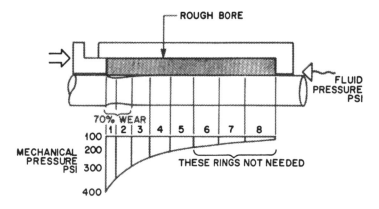

FIG. 11.14.   Pressure distribution through packed box seal.

TABLE 11.2 Modern Packing Materials

Packing	Properties
Graphite foil	High speed; high temperature; low leakage; infrequent adjustment
Aramid	Tolerant of abrasive and crystalline service; hard sleeve mandatory; higher leakage: sensitive to overtightening
PTFE/graphite	Inert; good thermal conductivity; material quality control critical

Packing ring size (section) is based on empirical rules for resilience (larger sections are more resilient, and are thus better at accommodating changes in service conditions) and ease of repacking. Flitney observes that a recent trend toward smaller section rings appears to lack supporting research and is thus aimed at cost reduction alone.

Packed box seal performance is dependent upon sleeve or shaft surface hardness, finish, and runout. All three affect packing life. Runout affects leakage, and Flitney reports that leakage varies as the cube of the runout.

With the development of better materials, a process accelerated by the need to eliminate asbestos, soft packing is regaining ground as a viable shaft seal. Karassik [11.2] argues that packing is the simplest seal and should therefore be used whenever possible. Pearse [8.5], speaking for mineral processing, observes that the packed box seal is considered cost effective, particularly when the suction pressure exceeds 15% of the pump pressure rise. Flitney reports the information in Table 11.2 for three modern packing materials.

Of the older packing materials, lead foil is still a good choice for high speed water service such as boiler feed.

Most packing runs on a sleeve, as shown in Fig. 11.9. While a separate sleeve makes concentricity harder to maintain, it does allow a replaceable working surface of the required hardness. Soft sleeves are bronze, iron, or 316 stainless steel. Hard sleeves are 13 chrome (450 BHN minimum) or hard-coated 316 stainless steel or a higher alloy. Common hard coatings are Colmonoy (fused), tungsten carbide, and chrome oxide.

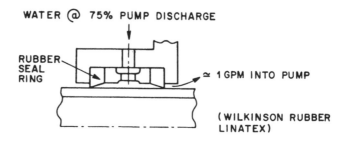

FIG. 11.15. Hydrostatic gland—diagrammatic (Wilkinson Rubber Linatex).

11.4. Fixed Packing—Rotating Shaft

Fixed packing for rotating shafts is a recent innovation, born of the need to seal solids-laden liquids with minimum dilution. Pearse [8.5] reports three recent devices:

Hydrostatic Gland. Two feather-edged rubber seal rings are separated by a seal cage (Fig. 11.15). Water at approximately 75% of pump discharge pressure is applied between the seals. Flow into the pump is typically 1 gal/min.

Simmering Seal. Similar to the hydrostatic gland, but using lip seals and having oil circulated between the seals to exclude the pumped liquid and lubricate and cool the seals.

Liquidyne Seal. Three sealing elements, retained in a four-piece bolted housing, run on a sleeve (Fig. 11.16). Clean flush water is injected into the

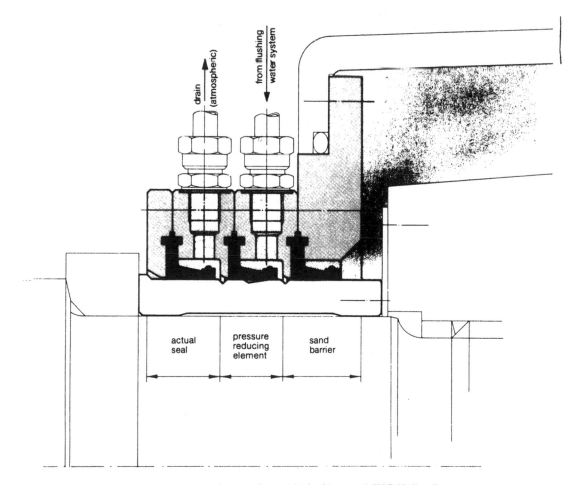

FIG. 11.16. IHC Liquidyne seal (IHC Holland).

seal, part of the flow passing under the sand barrier to exclude the pumped liquid. The balance of the flush water passes under the pressure reducing element, aided by grooves in the sleeve, and is returned to the water system or an atmospheric drain. Return line pressure drop determines the pressure drop across the outer, or actual, seal. The design has been tested to 435 lb/in.^{2}gauge; surface speed was not stated.

11.5. Hydrodynamic Seal

In their usual form, hydrodynamic seals are applied to overhung impellers. On the shaft side of the impeller, a "sealing impeller" or "expeller" (Fig. 11.17) acts to produce a pressure below atmospheric at the shaft opening. The result is a stable liquid/gas interface part way down the expeller vanes. Figure 11.17 shows the pressure profiles through a typical expeller sealed pump, operating with a positive suction.

A second, and still novel, design has a rotating, vaned housing which gathers leakage and accelerates it to a high peripheral velocity (Fig. 11.18). A stationary Pitot tube conducts the higher energy liquid back to the pump.

By definition, hydrodynamic seals are only effective while the pump is running. When the pump is shut down (or running at low speed), some form of auxiliary seal is necessary. Soft packing, fixed packing, V rings, and pneumatic bushings (inflated upon shutting down the pump's driver) are used for this function.

The effectiveness of a hydrodynamic seal depends upon its ability to develop, in conjunction with impeller pump-out vanes, a differential pressure greater than the maximum static pressure rise across the impeller. For expeller seals this limits the suction pressure to approximately 15% of the

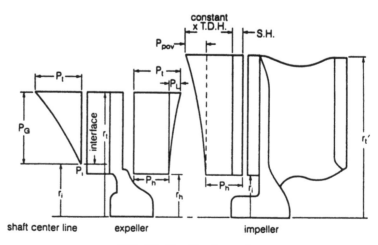

FIG. 11.17. Expeller seal.

pump's differential pressure. Pitot tube seals develop higher pressures, ranging from 55 to 95 lb/in.^{2}gauge at 3600 rev/min.

Because hydrodynamic seals do not rely on close clearances, they are tolerant of solids-laden liquids. With this capability they eliminate the need for special flush or quench arrangements. Expeller seals are more tolerant and are routinely used in slurry pumps. McDonald [11.3] reports quite extensive use of expeller seals in difficult chemical services. Utex [11.4] claims Pitot tube seals can handle liquids containing up to 35% solids.

Hydrodynamic sealing is not free; pumps so equipped absorb more power than those with conventional seals. Setting aside the question of energy consumption, since it probably nets out even in difficult services, the only

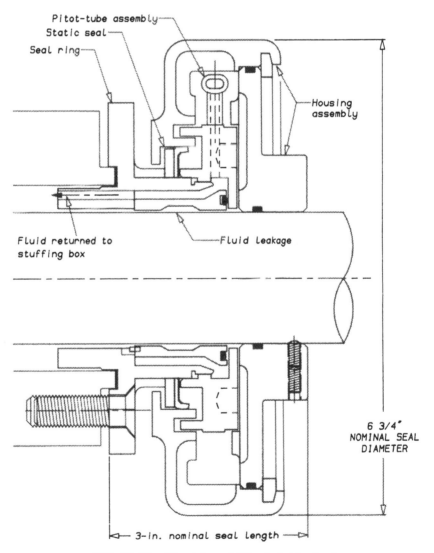

FIG. 11.18. Pitot tube seal (Utex Industries, Inc.).

problem the extra power poses is that of temperature rise in volatile liquids. The simplest means of overcoming this is to arrange for some circulation within the pump.

11.6. Axial Face (Mechanical) Seals

By arranging for the primary seal to be between a pair of contacting axial faces, one flexibly mounted to maintain contact (Fig. 11.19), the seal design no longer has to provide a resilient interface to accommodate minor rotor runout and radial movement. Free of this limitation, axial face seals can be designed for greater sealing capability and significantly lower leakage than soft or fixed packing. These two advantages have enabled more effective process design (higher pressures and temperatures), resulting in broad usage of such seals for chemical processing.

Along with more effective sealing, axial face seals compensate automatically for wear. This confers a third advantage, lower maintenance, since seal effectiveness is not dependent upon periodic adjustment.

Axial face or mechanical seals are not new. The design shown in Fig. 11.19 was patented in 1915 and incorporates features just now regaining popularity.

To realize acceptable service lives, the sealing interface in mechanical seals must be lubricated. Usually the lubrication is provided by the passage of a small amount of liquid, or gas in appropriate designs, through the interface. As a general rule, the leakage increases with sealing capability, but for most applications it is so low that it is not obvious. In this connection, Davidson [1.5] cautions on the need for adequate ventilation when the leakage is volatile. If the lubricating leakage ceases for reasons of pressure at the seal,

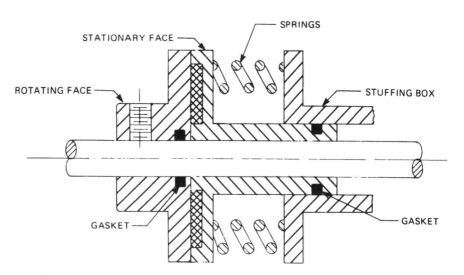

FIG. 11.19.   Mechanical seal patent issued in 1915 (BW/IP International, Inc.).

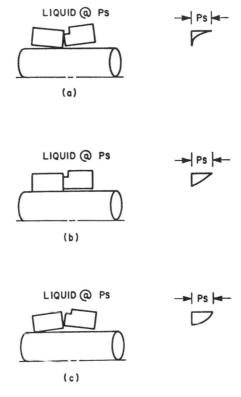

FIG. 11.20. Seal face orientation (exaggerated) versus pressure drop characteristic: (a) faces divergent—concave pressure drop; (b) faces parallel—linear pressure drop; (c) faces convergent—convex pressure drop.

temperature at the seal, or distortion (mechanical or thermal) of the faces, intimate contact between the faces occurs, which usually causes rapid wear.

Leakage through the seal interface is associated with a pressure drop across it. Seal leakage and seal face life are affected significantly by the characteristic of this pressure drop. Three basic pressure distributions are now recognized. Figure 11.20 shows the pressure distributions and the seal face orientations necessary to produce them.

To appreciate the significance of the seal face pressure distribution, it is necessary to consider net face loading and vaporization between the faces.

Net face loading is the resultant of hydrostatic force, flexible element force, and face separating force. Figure 11.21 illustrates this for both unbalanced and balanced seals. The face pressure distributions shown in Fig. 11.21 are linear, the usual first assumption. Should the faces actually be divergent, Fig. 11.20(a), with a concave pressure drop, the face separating force will be lower, hence the net face loading higher. At some point the face loading will exceed capacity, resulting in intimate contact and rapid wear. In the other direction, convergent faces with a convex pressure drop, Fig. 11.20(c), will result in a lower net face load. Up to a point this is desirable, but it does increase leakage and at the extreme can cause the seal to become unstable and "blow open;" see Wallace and David [11.5].

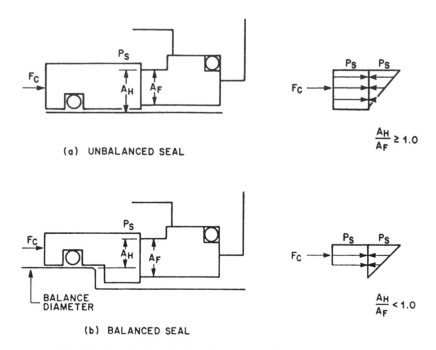

(a) UNBALANCED SEAL

(b) BALANCED SEAL

FIG. 11.21. Seal face loading: (a) unbalanced seal; (b) balanced seal.

Liquid vaporization part way across the interface lowers the face load capacity. If the vaporization occurs close enough to the face o.d. due to liquid temperature or face orientation or both, the face load capacity can fall to the point where rapid wear sets in. Figure 11.22(a) illustrates the problem. Three approaches are used to try to avoid it.

The first (Fig. 11.22b) involves raising the pressure at the seal or cooling the liquid or both to insure vaporization close to the face i.d., thus providing nearly full face load capacity. Wallace and David stress that the bulk temperature rise margin below vaporization is the important criterion (Fig. 11.23) and frequently requires more than the 25 lb/in.^{2} margin over the suction pressure mandated by API-610 [6.24].

As an alternative to rearranging the pump for higher pressure at the seal, the second approach is a seal designed to develop an effective liquid film between the faces by hydrodynamic action. Such designs usually have grooves of some form part way across one of the faces to promote the hydrodynamic action.

The third and most effective approach is also the simplest: avoid any vaporization by insuring the temperature of the leakage is below its atmospheric boiling point (Fig. 11.22c). While nominally simple, this approach is generally more expensive, thus it is often put aside in favor of the two lower cost alternatives. The combination of no vaporization and convergent seal faces (Fig. 11.20c), realizes the maximum capability of axial face seals.

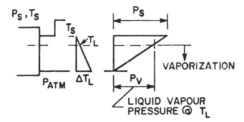

a) VAPORIZATION CLOSE TO SEAL FACE O.D.

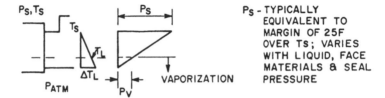

b) VAPORIZATION CLOSE TO SEAL FACE I.D.

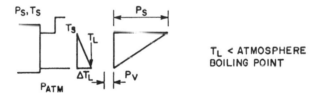

c) NO VAPORIZATION ACROSS SEAL FACE

FIG. 11.22. Seal face performance—point of vaporization versus sealed pressure and temperature: (a) vaporization close to face o.d.; (b) vaporization close to face i.d.; (c) no vaporization across seal faces.

Figure 11.22 shows a certain temperature rise, ΔT_L, in the seal leakage. For a given liquid, this temperature rise depends upon the friction heat generated by the seal and how well that heat is dissipated. The heat generated is a function of face loading and the coefficient of friction, the latter dependent upon the material combination and lubrication. Heat dissipation is a function of thermal conductivity through the faces and transfer to the sealed liquid, the usual heat sink. Solving all this is the seal designer's problem; sufficient here to just note the broad importance of face materials.

Axial face seal capability, categorized by the three methods of face lubrication, is given in Fig. 11.24. Within this general PV chart there are specific limits, either as PV or in the form of Fig. 11.23, for various face materials and liquids. These data are available from the seal manufacturers.

Mechanical seal designs now include many variations. The one clear distinction between them is the means of sealing the flexibly mounted face. This is a fundamental problem, requiring an arrangement which seals effectively without impairing the dynamic behavior of the face. Two basic designs

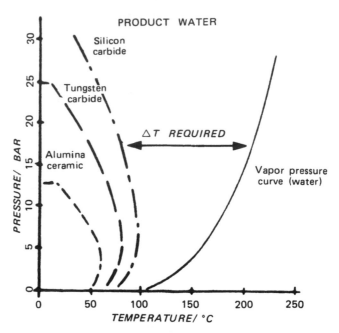

FIG. 11.23. Typical curve showing temperature margin requirement for various face material combinations (Flexibox International).

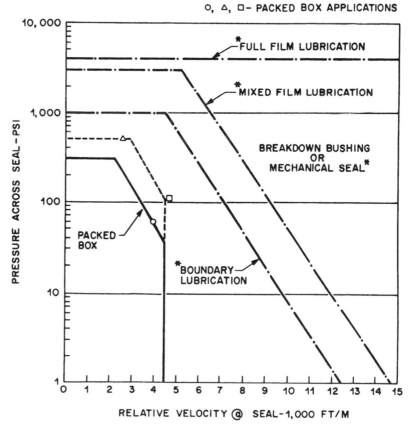

FIG. 11.24. Shaft seal capabilities (after Durametallic).

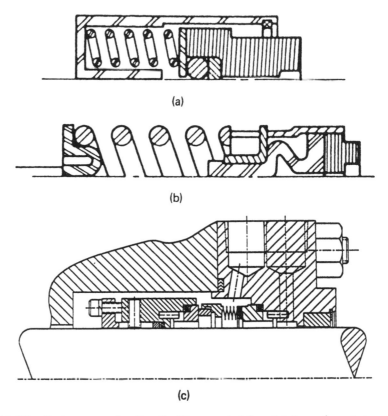

FIG. 11.25. Basic means of sealing flexibly mounted face: (a) dynamic gasket—rotating (Houdaille John Crane); (b) elastomer bellows—rotating (Houdaille John Crane); (c) metal bellows—stationary (EG&G Sealol, Inc.).

are used: some form of dynamic gasket (Fig. 11.25a) or a bellows, either elastomer (Fig. 11.25b), polymer, or metal (Fig. 11.25c). Seals using a dynamic gasket are often referred to as "pusher" seals, since the dynamic gasket is "pushed" along the sleeve to compensate for face wear.

Figure 11.26 shows a mechanical seal design classification based on the distinction in flexible face sealing. Table 11.3 provides comparable data for the two approaches.

Compounding the variations in design, mechanical seals are used in a number of arrangements. Those in common use are shown in Fig. 11.27.

Mechanical seal design and arrangements cannot be dealt with adequately in this volume. See Netzel [in Ref. 1.1], Mayer [11.6], or manuals by the major manufacturers for detailed treatments.

Despite what is known of mechanical seal behavior, the age of the concept, and their demonstrated general reliability, mechanical seals are still considered the most unreliable element in process pumping. Byrd [6.20], drawing on data compiled from 1969 through 1972, reports an average seal life of 89 days. Wallace and David cite data showing that of the losses caused by pump fires, 54% were attributable to seal failures. Martel et al. [11.7] report a survey showing 69% of unscheduled pump outages were caused by seal failures. An

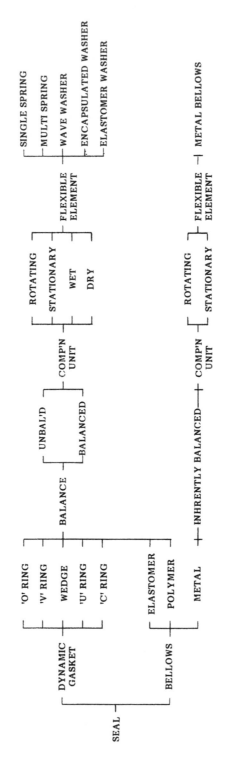

FIG. 11.26. Mechanical seal design classification.

TABLE 11.3 Comparison of Secondary Seals[a]

Secondary Seal	Dynamic		Bellows	
	Gasket O ring V, U, C ring Wedge	Elastomer Polymer		Metal
Compression	Single or multiple springs			Bellows
Size range	Broad 0.5 to 20 in.	Broad		Narrower 0.75 to 5.00 in.
Pressure limit	High To 3000 lb/in.^{2}gauge	Low To 30 lb/in.^{2}gauge[b]		Intermediate To 350 lb/in.^{2}gauge
Temperature limit	Intermediate −450 to 300°F[b]	Low 0 to 120°F		High −450 to 800°F
Balance	Available	Available		Inherent
Hysterisis	Moderate		Negligible	
Risk of fretting	High[c]	None		None
Special designs	Available	Not available		Limited
Special materials	Available	Not available		Limited
Number of parts	More	More		Fewer
Cost				Usually higher

[a]After Durametallic.
[b]Limits per Wong [11.9].
[c]Reduced with stationary compression unit.

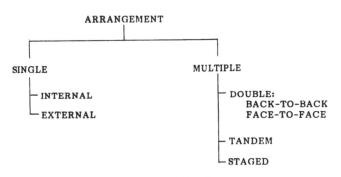

FIG. 11.27. Mechanical seal arrangements.

TABLE 11.4 Failure in Seal Components

	% of Incidents	
Problem	Martel et al. [7]	Adams [8][a]
Face wear (all causes)	45	51
Dynamic gasket	40	N.A. (bellows)
Static gasket	5	16
Corrosion	N.R.	18
Other	10	15

[a]Excluding problems caused by pump maloperation.

indication of the seal components most susceptible to failure is given in Table 11.4 by Martel et al. and Adams [11.8], the latter for seals running above 350°F.

Improving seal reliability involves many corrective measures, some fundamental but most detailed. Those measures deemed fundamental are outlined below.

1. Seal Housing. Most mechanical seal designs are based on housing dimensions suitable for soft packing, a legacy of older pump designs and the desire to be able to interchange packing or a mechanical seal. Wong [11.9] and Block, [11.10] stress the need to recognize that better mechanical seal designs require more radial space than packing. Wong adds that in the rare instances where retrofitting a mechanical seal with packing is feasible, a simple insert is sufficient. Proposed revisions of ANSI B73.1 and 2 and API-610 are considering mandating more radial room for mechanical seals.

Associated with providing more radial space for mechanical seals, there is evidence that an open, internally flushed seal housing of the form shown in Fig. 11.28 offers several advantages. Byrd [6.20] reports elasto-

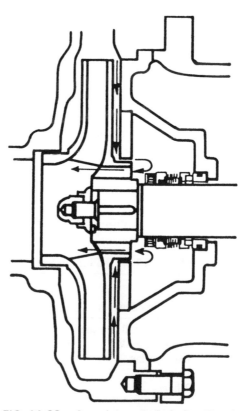

FIG. 11.28. Open, internally flushed seal housing.

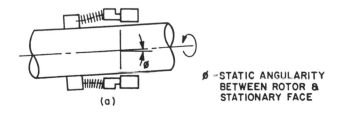

ϕ —STATIC ANGULARITY
BETWEEN ROTOR &
STATIONARY FACE

(a)

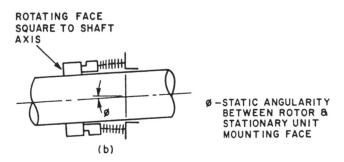

ROTATING FACE
SQUARE TO SHAFT
AXIS

ϕ —STATIC ANGULARITY
BETWEEN ROTOR &
STATIONARY UNIT
MOUNTING FACE

(b)

FIG. 11.29. Compression unit tolerance of angularity.

mer bellows seals in such housing tolerating "very high concentrations of dirt, swarf, sand, solids, and gas." Pearse [8.5], discussing tests of mechanically sealed slurry pumps, notes the superiority of open seal housings, citing good cooling and lubrication qualities, hence lower temperature rise at the seal and minimum air entrapment.

2. Compression Unit. The mechanical seal patented in 1915 (Fig. 11.19) has a dry, stationary compression unit. This arrangement confers two advantages. First, the compression unit parts are not subject to corrosion, fouling (clogging, scaling), or erosion by the pumped liquid. Second, being stationary, the compression unit is not subject to the dynamic effects of rotation and is tolerant of shaft angularity through the seal. Freedom from the dynamic effects of rotation is mandatory above 4500 ft/s (Bloch [11.10]; spring distortion and unbalance) or 575°F (Wong [11.9]; bellows creep). Tolerance of angularity is a product of the compression unit not having to cycle to accommodate static angular misalignment (see Fig. 11.29). The rotating face must, of course, be square to the shaft axis to realize this.

Metal bellows seals have a wet compression unit whether stationary or rotating. Adams and Bloch note the self-cleaning effect of a rotating bellows, and Adams recommends a rotating bellows when the solids exceed 2% by weight. An open seal housing with high, swirling circulation might provide sufficient cleaning when other circumstances require a stationary metal bellows. Byrd reports good results with rotating, internal

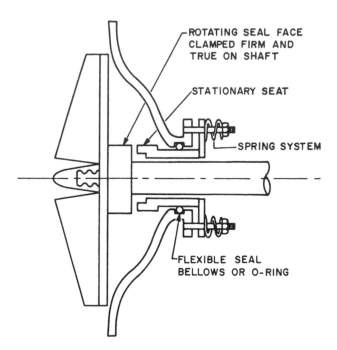

FIG. 11.30. Chemical pump seal proposed by Byrd [6.20].

elastomer bellows seals, but cautions against external PTFE bellows seals, claiming they are susceptible to solids being centrifuged into the seal interface.

3. Dynamic Gasket. Dynamic gaskets are prone to fretting damage, particularly when made of PTFE and used to seal a nonlubricating or a slightly abrasive liquid. As noted earlier, a bellows seal avoids this but cannot always be used due to pressure limits and cost. By using a stationary compression unit, see Item 2 above, the tendency to dynamic gasket

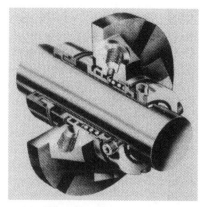

FIG. 11.31. Slurry seal with stationary, external compression unit (Flexibox International).

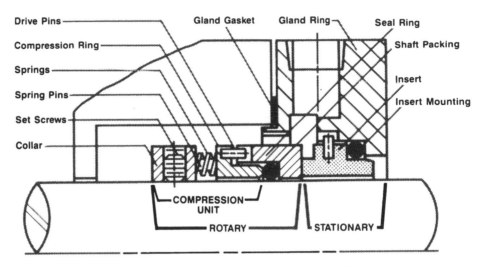

Drive Pins — Gland Gasket — Gland Ring — Seal Ring

Compression Ring — Shaft Packing

Springs —

Spring Pins — Insert

Set Screws — Insert Mounting

Collar —

COMPRESSION UNIT

ROTARY — STATIONARY

FIG. 11.32. Direct mounted mechanical seal (from the *Dura Seal Manual*, Copyright 1987, Durametallic Corporation).

fretting is reduced. A second, notable benefit is the ability to arrange the dynamic gasket so it passes over a clean surface to compensate for face wear. Byrd advocates a chemical pump seal (Fig. 11.30) incorporating this arrangement along with an adjustable compression unit and an open seal housing. Proprietary designs for severe service, such as Fig. 11.31, are approaching this.

4. Seal Mounting. Seals can be direct or cartridge mounted. Direct mounting (Fig. 11.32) involves installing the seal as two or three separate pieces, then either setting it *in situ* or relying upon the pump's dimensions to determine the setting. Cartridge mounting (Fig. 11.31) has the seal installed as an assembled, pre-set unit, thus reducing the risk of damage, contamination, or incorrect setting. Not by any means a new idea, cartridge mounting has gained in acceptance as its cost effectiveness has become more apparent. Cartridge seals should be installed directly on the shaft. Installing them over an intermediate sleeve adds a leak path and makes it harder to maintain concentricity.

5. Arrangement. Single, external seals are pressure limited, susceptible to solids (see Item 2 above), and all but superseded by designs such as shown in Fig. 11.31.

Double seals (Fig. 11.33) were the recommendation for solids-laden, flammable, or toxic liquids. The arrangement, however, is complicated and fraught with problems:

Wallace and Davis [11.5] report that the final installed cost of a double seal is some 50 times that of a simple single seal.

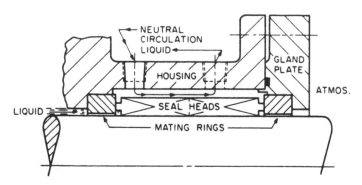

FIG. 11.33. Double seal, back-to-back arrangement (Houdaille John Crane).

Some pumped liquids cannot tolerate the low dilution from barrier liquid
leakage across the inner seal (Byrd [6.20], Shepherd [11.11]).

Despite a higher pressure barrier liquid, light hydrocarbons can migrate
through the interface and accumulate in the barrier liquid system
(Wallace and David) [11.5]).

Fine solids have demonstrated a tendency to accumulate in the region
between the inner face and the rotor, eventually closing the clearance
and causing fracture of the face.

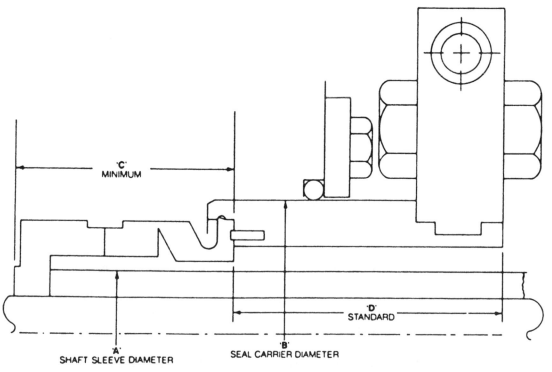

FIG. 11.34. Elastomer energized slurry seal (Type RIS, BW/IP International, Inc.).

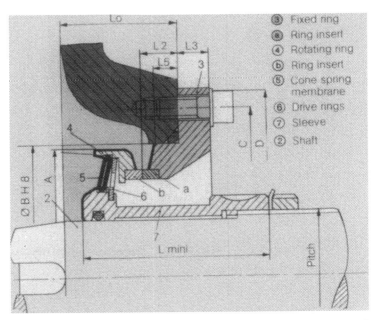

FIG. 11.35. Slurry seal energized by encapsulated Belleville spring (Le Carbone-Lorraine, Cefilac Etancheite).

Alternatives being used instead of double seals are:

Single seals specially designed for solids laden liquids (Figs. 11.31, 11.34, and 11.35). See notes on face materials in Item 6 below.

Single seal with a quench, bushing or lip seal retained, in applications where leakage dilution is sufficient to protect the environment.

Single seals with a backup seal. The backup seal can be a gas seal running in air (Fig. 11.36) or an abeyant seal (Fig. 11.37) whose faces are brought into contact by high leakage.

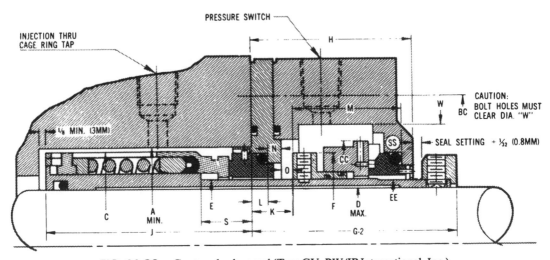

FIG. 11.36. Gas-type backup seal (Type GU, BW/IP International, Inc.).

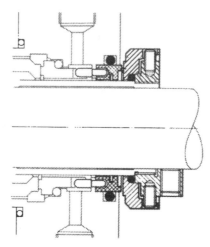

FIG. 11.37.   Abeyant backup seal (Flexibox International).

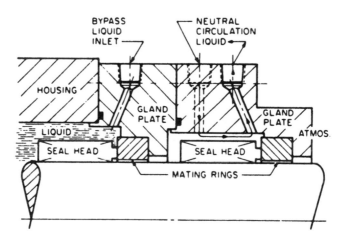

FIG. 11.38.   Tandem seals (Houdaille John Crane).

TABLE 11.5   Relative Lives for Face Materials

Face Materials	Liquid	Relative Life of Silicon Carbide vs Carbon Faces
Stellite vs C	HC	3
Stellite vs C	Water	7
WC vs C	All	1.5

Tandem seals (Fig. 11.38) in which both seals are of equal capability and an API plan 52 system lubricates the secondary seal and monitors primary seal leakage. The final installed cost of a tandem seal is considered to be 10 times that of a simple single seal (Wallace and David [11.5]).

Hermetically sealed pumps, see Chapter 6, when product leakage to the atmosphere, or atmospheric contamination of the product, cannot be tolerated.

In some circumstances, legislation can mandate double seals for various liquids. Should that happen, the legislation has to prevail.

6. Face Materials. Of the face material combinations available, silicon carbide versus carbon offers low friction plus high thermal conductivity, hence a low temperature rise in the interface leakage. One manufacturer submitted the information of Table 11.5 for typical relative lives for various face material combinations. Byrd notes that carbon vs carbon and alumina vs PTFE faces have good dry running properties. Thus they are desirable materials when other conditions allow.

   To aid heat dissipation, hence reduce the risk of vaporization across the seal interface, the usual practice is to rotate the better thermal conductor (Bloch [11.10]). Wallace and David [11.5] report difficulties with light HC seals arranged this way. They attribute the difficulty to vapor accumulation (centrifuge action) and recommend making the better conductor stationary for such services.

7. Selection. Systematic selection of mechanical seals avoids oversight, aids troubleshooting, and enables refinement of the selection process. Block and Wong advance selection systems based on a conservative assessment of published seal capabilities and operating experience. Establishing the actual or necessary seal environment, see Section 11.1, is, of course, fundamental to a correct solution.

11.7. Reciprocating Pump Seals

Seals for reciprocating pumps can be classified as inside or outside (see Fig. 11.39 and 11.40). By the definition given in Chapter 8, those pumps with an inside seal are classified as piston pumps. The seals for piston pumps are dealt with in Chapter 8. This section deals only with stationary outside seals.

All the seals used in reciprocating pumps are close radial clearance devices whose operation depends upon a certain degree of leakage, the degree depending upon the seal type and service. In broad terms the leakage rate is proportional to the circumference of the leakage path for a given pressure, and inversely proportional to the length of the seal and liquid viscosity.

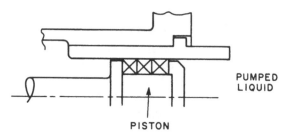

FIG. 11.39.   Inside (moving) seal.

In high pressure plunger pumps the predicted rate of plunger/seal wear can be a determining factor in pump speed, hence pump size. Davidson [1.5] provides a means of making this assessment. The assessment relies upon *PV*, pressure times mean velocity, being approximately constant above certain threshold values (0.7 ft/s and 175 lb/in.²gauge). By using this relationship, an empirical equation for allowable wear and data for the wear rate of various plunger materials in various liquids, an acceptable plunger speed can be determined.

Three forms of seal are used in reciprocating pumps:

Soft, square section packing
Formed rings, e.g., V rings, U cups, etc.
Floating rings

Seals are installed in one of three ways: adjustable, self-adjusting (spring loaded), and fixed. Table 11.6 shows the installation arrangements commonly used with the various seal forms.

An adjustable installation (Fig. 11.40) employs a gland to apply axial compression to the seal. Two gland designs are used: single thread, either internal or external (Fig. 11.40 shows an external thread), and multiple stud

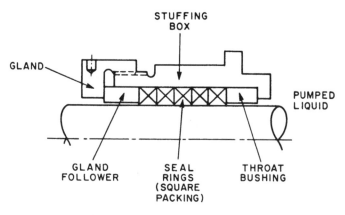

FIG. 11.40.   Outside (stationary) seal.

TABLE 11.6   Installation Arrangements

Seal Type	Installation		
	Adjustable	Self-Adjusting	Fixed
Soft packing	×	×	—
Formed rings	×	×	×
Floating rings	—	—	×

(usually 2). The prime disadvantages of an adjustable installation are uneven loading of the sealing elements and the need for periodic adjustment.

By replacing the adjustable gland with some form of flexible element, such as a spring, it is possible to improve the loading on the sealing elements and eliminate the need for periodic adjustment. Figure 11.41 shows a self-adjusting installation of V rings. An important aspect of self-adjusting designs is the spring constant of the flexible element. To maintain a relatively constant load on the seal elements as they wear, the spring constant needs to be low. Properly designed, self-adjusting seals are the reciprocating pump's equivalent of the mechanical seal; see Henshaw [in Ref. 1.3] and Coopey [11.12].

When it is considered desirable to avoid any degree of unequal loading in the sealing elements, a fixed seal installation is employed. With this arrangement the whole seal assembly is clamped tight and each ring or element is able to act independently. Figure 11.42 shows a fixed installation of formed rings.

All forms and installations of reciprocating seals use throat bushings and followers. The clearance between these components and the rod or plunger has a major bearing on seal performance. Typical values are shown in Table 11.7.

Materials used for throat bushings and followers must have high resistance to galling. In this connection, Miller [11.13] notes that Niresist against Colmonoy is a poor combination.

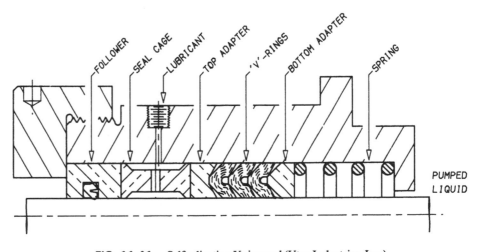

FIG. 11.41.   Self-adjusting V ring seal (Utex Industries, Inc.).

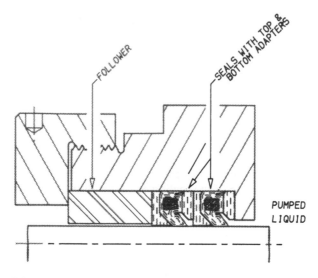

FIG. 11.42. Fixed (nonadjustable) seal (Utex Industries, Inc.).

Plunger or rod surface finish is critical to seal performance. A finish of 16 RMS is considered ideal. A rougher surface will damage the seal elements; a smoother surface may not carry sufficient lubricant into the region under the sealing elements.

Seal element section, at least for soft packing and formed rings, is typically 0.375 to 0.500 in. and should not be larger than 0.750 in.

In horizontal pumps the seal assembly often serves to help support the pumping element. This additional function imposes special requirements on the design and materials of the seal.

Stuffing box wear, known as "washboarding," is a common problem in reciprocating pump seals. The wear is a consequence of packing motion and is best avoided by maintaining sufficient packing load to prevent motion.

Lubrication of the seal elements is fundamental to performance. Broadly three systems are in general use:

Pump liquid leakage: Suitable when the pumped liquid has reasonable lubricating properties and its leakage to atmosphere can be tolerated.

TABLE 11.7 Throat Bushing and Follower Clearances (0.001 ins diam.)

Plunger Diameter (in.)	Pressure Drop (lb/in.^{2})		
	Under 500	500 to 3000	3000 and Over
3 and under	0.006	0.004	0.003
4 to 8	0.008	0.006	0.004
8 to 10	0.010	0.008	0.005
10 to 12	0.012	0.010	0.006

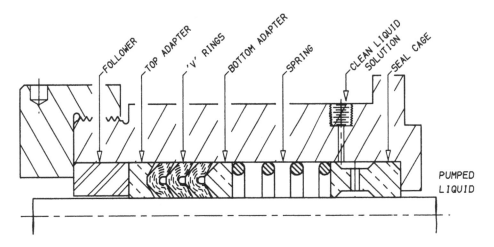

FIG. 11.43. Injected seal (Utex Industries, Inc.).

Drip lubrication: Oil, or a similar lubricant, is fed to either the atmospheric
  side of the seal or to a seal cage within the seal assembly. Figure 11.41 shows
  such a seal cage.

Injection: In corrosive or abrasive service, it is desirable to keep the pumped
  liquid from direct contact with the seal elements. This is achieved by
  injecting an inert liquid, compatible with the pumped liquid, through a
  close clearance bushing inboard of the seal elements. Figure 11.43 shows a
  seal so arranged. The injection can be either synchronized to the pumping
  element's suction stroke or at a constant pressure above the pump's
  discharge pressure. Synchronized low pressure injection is easier to provide, but is not as effective as continuous high pressure injection. The flow
  required is determined by the appropriate "flushing velocity" (see Section
  11.1) and the bushing annular area. As a rough estimate, use 2 to 3% of
  rated pump capacity.

In services where leakage of the pumped liquid is able to lubricate the
packing but leakage directly to atmosphere cannot be tolerated, a barrier
liquid system is employed. This functions by circulating an inert liquid
through the packing assembly to collect the primary seal leakage. The barrier
liquid system can be once through or recycle with means for separating the
primary seal leakage. Figure 11.44 shows a seal arranged for barrier liquid
circulation.

Soft Packing

To function effectively, soft packing has to be sufficiently resilient to seal at its
ID and OD, has to have a low coefficient of friction, high thermal conductivity, and low thermal expansion. Traditionally these requirements were met by
making a soft packing assembly of several materials. A typical example is lead

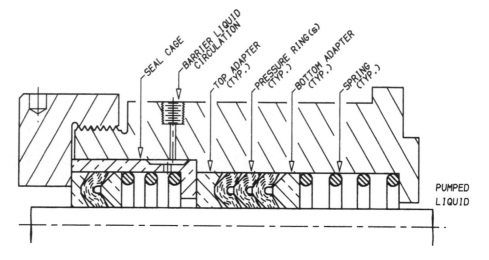

FIG. 11.44. Barrier liquid seal (Utex Industries, Inc.).

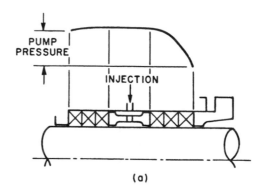

(a)

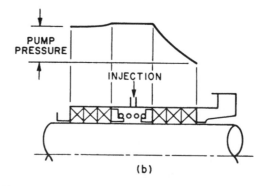

(b)

FIG. 11.45. Pressure distribution across soft packing.

foil rings for strength and low friction, braided flax rings for resilience, and reinforced phenolic followers for strength and bearing. PTFE offers low friction but has poor thermal conductivity and high thermal expansion, so is only used when its high corrosion resistance is needed. Carbon and graphite fibers, relatively recent developments, satisfy all the requirements and are thus being used with increasing frequency.

The principal limitation of soft packing in an adjustable installation is uneven distribution of loading and pressure drop (see Fig. 11.45a). A self-adjusting installation (Fig. 11.45b) helps even out the pressure distribution.

Of the sealing elements used in reciprocating pumps, soft packing generally has the shortest service life. To make replacement easier, soft packing is often made in a split ring form. When so constructed, the pressure drop that can be taken across the seal is limited.

Formed Rings

High pressure drop capability, flexibility in installation, and long service life make formed rings the most common form of seal in high pressure reciprocating pumps.

Under pressure and operating in one direction, formed rings are virtually leak free. When subjected to pressure cycles and alternating motion, as in a reciprocating pump, cyclic distortion of the rings causes a small amount of leakage. The amount of leakage depends upon the extent of distortion and the frequency of cycling.

Formed rings for "outside" pump seals are balanced, i.e., they have two sealing edges. Rings termed unbalanced (see Fig. 11.46) have only one sealing edge and are commonly used as "inside" seals in reciprocating pumps.

A single U seal can contain up to 15,000 lb/in.^{2} gauge over a temperature range of -70 through 450°F. Such a seal offers the advantage of simplicity. Its drawback is catastrophic failure once leakage starts. Multiple V rings can be arranged into a shorter assembly, are capable of sealing up to 30,000 lb/in.^{2} gauge, and tend to fail progressively. If arranged for adjustable installation, it is possible to "take up" leakage for a limited time. Figure 11.47 shows a single U seal, and Fig. 11.41 shows a multiple V ring seal. Formed

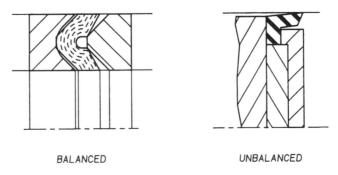

BALANCED   UNBALANCED

FIG. 11.46. Seal ring balance (Utex Industries, Inc.).

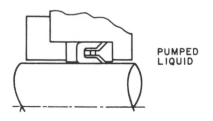

FIG. 11.47.   Single U seal (after Smith [11.14]).

rings must be supported by adaptors (see Fig. 11.41). Adaptors can be metal or some resilient material. Metal offers low distortion; resilient material a greater tolerance of dimensional errors.

Seal ring material has to match the design pressure drop. The ideal material is soft and resilient in order to give a low coefficient of friction. Provided the material's strength is sufficient to prevent extrusion into the follower clearance, this will be the case. Once extrusion occurs, however, the coefficient of friction will increase significantly. Typical ring materials are various compounds of rubber (of a composition suitable for the sealed liquid) and PTFE reinforced with fabric.

While formed rings can be installed in all three arrangements, see Table 11.6, the most common arrangement is self-adjusting. Figure 11.41 shows such an arrangement. An adjustable arrangement, when used, is similar to Fig. 11.48. A fixed arrangement, in which the rings are clamped tight, is shown in Fig. 11.42. The installation of such arrangements should carefully follow the manufacturer's recommendations.

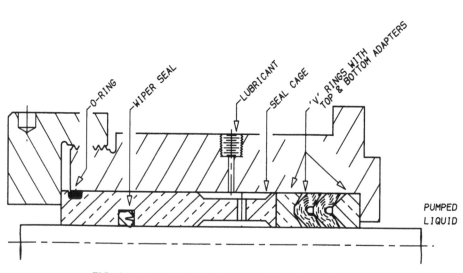

FIG. 11.48.   Adjustable V ring seal (Utex Industries, Inc.).

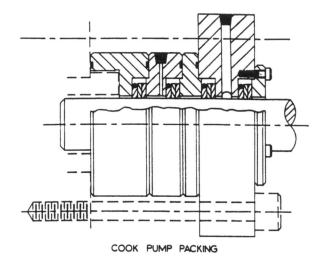

COOK PUMP PACKING

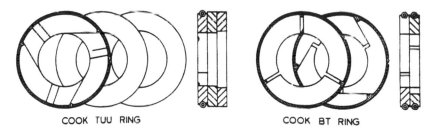

COOK TUU RING   COOK BT RING

FIG. 11.49. Floating ring seal (C. Lee Cook, Louisville, Kentucky).

Floating Ring

The essential feature of this seal arrangement is a series of seal ring assemblies, each located in a separate groove (see Fig. 11.49) with the objective of realizing a progressive pressure drop over the seal. From work reported by Scheuber [11.15], the pressure drop per ring will likely vary with wear instead of being equal. Each seal ring assembly typically consists of two or more rings, segmented or solid, depending upon the detail design (see Fig. 11.49). Each ring is free to move radially, and thus finds its optimum position. The rings can be produced in a wide variety of materials to better handle severe services. By design, the clearance in floating ring seals is greater than in formed rings, and thus the leakage is higher. Given their advantages and disadvantages, floating ring seals are limited to severe service such as abrasives, ammonia, and high temperature liquids at 2000 lb/in.^{2}gauge and higher.

References

11.1. R. K. Flitney, "Soft Packings," *Tribol. Int.*, *19*(4), 181–183 (August 1986).

11.2. I. J. Karassik, "To Seal or to Pack," *Plant Serv.*, pp. 18–21 (July 1987).

11.3. D. P. McDonald, "Hydrodynamic Seals," *World Pumps*, 1974, pp. 110–114.

11.4. Utex Industries, Inc., "Seal Handles Solids-Containing Fluids, Eliminates Leakage," *Chem. Eng.*, p. 33 (February 17, 1986).

11.5. N. M. Wallace and J. David, "The Design and Application of Seals for Light HC Service," in *Proc. BPMA 9th Int. Tech. Conf., UK*, April 1985, pp. 241–260.

11.6. E. Mayer, *Mechanical Seals*, Newnes-Butterworths, London, 1977.

11.7. Y. Martel et al., "Reduce Costs with Metal Bellows Shaft Seals," *Hydrocarbon Process.*, pp. 39, 40 (October 1987).

11.8. W. V. Adams, "Better High Temperature Sealing," *Hydrocarbon Process.*, pp. 53–59 (January 1987).

11.9. W. Wong, "Seal Engineering for Reliability," in *Proc. BPMA 9th Int. Tech. Conf., UK*, April 1985, pp. 223–240.

11.10. H. P. Bloch, "Selection Strategy for Mechanical Shaft Seals in Petrochemical Plants," in *1st Int. Pump Symp., Houston, Texas*, May 1984, pp. 115–121.

11.11. K. Shepherd, "Sealing in the Petrochemical Industry," *Lubr. Eng.*, *36*(1), 40–44 (January 1980).

11.12. W. Coopey, "A Fresh Look at Spring Loaded Packing," *Chem. Eng.*, pp. 278, 280, 282, 284 (November 6, 1967).

11.13. J. E. Miller, *Packing Notebook*, Unpublished Compilation.

11.14. J. N. Smith, "Theory and Design of Large Diameter Packings," in *Proc. Nat. Conf. Industrial Hydraulics*, pp. 121–130.

11.15. K. Scheuber, "Dynamic Pressure Distribution in the Cylinder Packing of Compressors for Very High Pressures," *Sulzer Tech. Rev.*, (3), 100–104 (1980).

12. Couplings

Most pumps have a separate driver. To transmit torque from the driver to the pump, their shafts are generally connected with a coupling. In some arrangements the coupling also provides support for the rotor of one of the coupled machines. Figure 12.1 shows the three possible basic arrangements.

Given perfect alignment between the coupled machines, the couplings for 4 and 3 bearing arrangements can be rigid (couplings for 2 bearing arrangements are of necessity rigid). Perfect or very precise alignment is extremely difficult to achieve and to maintain, thus the couplings used for 4 and 3 bearing arrangement pumps are flexible.

A flexible coupling is able to tolerate a small amount of misalignment between the machines' axes of rotation. Flexible couplings are not universal

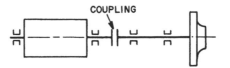

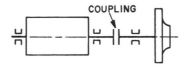

**(a) 4 BEARING ARRANGEMENT - COUPLING
        TRANSMITS TORQUE ONLY**

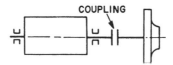

**(b) 3 BEARING ARRANGEMENT - COUPLING
        TRANSMITS TORQUE & RADIAL REACTION**

**(c) 2 BEARING ARRANGEMENT - COUPLING
        TRANSMITS TORQUE & BENDING MOMENT**

FIG. 12.1. Three basic driver/pump rotor arrangements and consequent coupling loads.

joints and will not give satisfactory service unless well aligned. How well aligned depends upon the coupling type and the machine service conditions.

Misalignment takes two forms: parallel and angular. Figure 12.2 illustrates the distinction. Table 12.1 summarizes the type of misalignment tolerance necessary for the three basic machine arrangements.

A wide variety of flexible coupling designs is used for pump drives. The first distinction is between materials. Many designs employ elastomer or

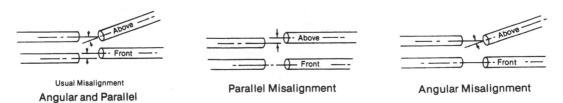

FIG. 12.2. Misalignment can be measured as parallel and angular offset (Courtesy Rexnord).

TABLE 12.1 Misalignment Tolerance Necessary

	Misalignment Capability	
Machine Arrangement	Parallel	Angular
4 bearing	×	×
3 bearing	—	×
2 bearing	—	—

polymer elements to realize flexibility; the remainder employ various all-metal configurations. The more usual coupling designs are listed below.

Material	*Types*
Elastomer	Block, ring, sleeve
All-metal	Disc, diaphragm, spring grid, gear

The virtue of elastomer couplings is freedom from the need for periodic lubrication. Because torque is transmitted through a low strength material, couplings employing elastomer elements are larger than equivalent all metal couplings. One consequence of this is that elastomer couplings become relatively more expensive than all-metal couplings as size increases. They also become heavier, which can present problems of high rotor inertia and overhung weight.

Elastomer block couplings, actually either blocks or bushings (Figure 12.3), transmit torque by compression of their elastomer elements. Flexure of

FIG. 12.3. Elastomer block flexible coupling (Kop-Flex, Inc.).

FIG. 12.4. Elastomer ring coupling (Lovejoy, Inc.).

these same elements provides misalignment capacity. The couplings are therefore torsionally "soft" but tend to be fairly stiff radially, thus their misalignment capacity is low. Some designs have available blocks of various hardness to permit "tuning" the coupling to avoid torsional resonance, a useful feature with pulsating drives.

Elastomer ring couplings (Fig. 12.4) transmit torque by compression in an alternately bolted ring. To transmit torque by compression the ring is installed compressed (banded) or compressed during installation (radial

FIG. 12.5. Sleeve-type clamped elastomer coupling (Dodge Manufacturing Division, Reliance Electric).

bolts). By working in compression, the elastomer ring avoids the need for reinforcement. Elastomer ring couplings are more tolerant of misalignment than elastomer block, but should be used in pairs if high misalignment is expected.

Elastomer sleeve couplings overcome the limited misalignment capacity of elastomer block and ring couplings. Torque is transmitted by shear in the elastomer sleeve. Using a sleeve as the flexible element confers the high misalignment capacity. The more usual arrangement (Fig. 12.5) uses a split, convex section sleeve. Such an arrangement realizes the highest torsional capacity for a given size and allows easy assembly, but does limit rotative speed. Two alternative arrangements for higher speeds are a diaphragm form sleeve (Fig. 12.6a) and a concave sleeve (Fig. 12.6b), both of which are continuous. Yet another approach is to retain a split sleeve but make it from a stiffer material. Polk [12.1] reports good experience with such couplings.

The second group of couplings, all-metal, offers the following advantages over those employing elastomer elements:

1. Smaller size for given capacity.
2. Lower cost for higher capacities.
3. Higher allowable rotative speed.
4. Greater tolerance of adverse environment, e.g., ultraviolet light, chemical, temperature.

(a)

(b)

FIG. 12.6.   (a) Continuous diaphragm-type elastomer coupling (Dodge Manufacturing Division, Reliance Electric). (b) Continuous concave sleeve-type elastomer coupling (Falk Corporation).

FIG. 12.7. Flexible metal disc coupling (Courtesy Rexnord).

All-metal couplings can be subdivided into two categories: lubricated and nonlubricated. Lubricated couplings realize flexibility by relative movement between coupling components. Nonlubricated couplings rely upon flexure of some form of flexible element. The virtues of nonlubricated couplings are freedom from periodic maintenance and lower forces from misalignment; see Gibbons [12.2].

Flexible metal disc couplings (Fig. 12.7) transmit torque by tension in an alternately bolted metal disc. To reduce stress the disc is usually a laminate of thin plates. A single disc unit has high angular flexibility, allowed by deflection of the disc, but very high radial stiffness. To realize parallel flexibility, two separated disc units are necessary. The coupling in Fig. 12.7 is so arranged.

Diaphragm couplings are similar to the disc type. The distinction is the form of flexible element and its effect on coupling size. Torque in diaphragm couplings is transmitted by shear in the diaphragm. The minimum diameter of the diaphragm is determined by allowable stress. Coupling flexibility requires that the diaphragm outside diameter be quite a deal larger than the inside diameter. These two factors combine to make diaphragm couplings larger in diameter than the equivalent disc coupling; see Mancuso [12.3] for further discussion. Some designs of diaphragm couplings make use of the high shear stress at the diaphragm inside diameter to provide a "shear pin" for overload protection. A typical diaphragm coupling is shown in Fig. 12.8.

Spring grid couplings (Fig. 12.9) employ a flexible metal element to transmit torque with a degree of torsional flexibility. Angular and parallel flexibility is realized by a combination of movement between the spring grid and the hub teeth and flexure of the grid itself. The movement within the coupling requires that the coupling be lubricated. Bending of the grid produces high radial loads, thus spring grid couplings need to be well aligned to avoid imposing high loads on the coupled machines' bearings.

Gear-type couplings offer the highest torque capacity for a given size, a product of their transmitting torque through gear teeth. As with disc and diaphragm couplings, a single gear element has only angular flexibility, realized by sliding between the teeth. Parallel flexibility therefore requires two

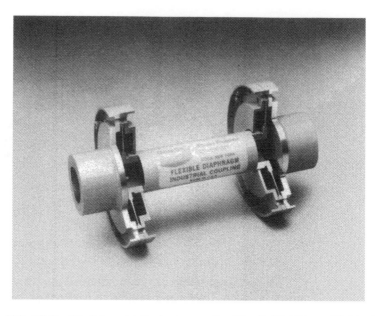

FIG. 12.8. Flexible metal diaphragm coupling (Bendix Fluid Power Division).

separated gear elements. Figure 12.10 shows a typical double engagement gear-type coupling. The misalignment capacity of gear-type couplings is limited by tooth sliding velocity. Wright [12.4] reports that a mean velocity of 5 to 8 ft/s is the upper limit for good service life. Because gear-type couplings realize flexibility by sliding between their teeth, lubrication is critical. Inade-

FIG. 12.9. Spring-grid coupling (Falk Corporation).

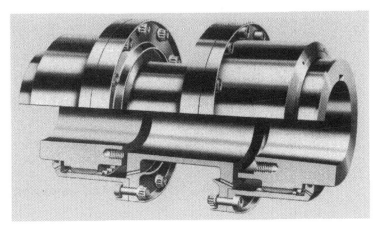

FIG. 12.10. Double engagement gear-type spacer coupling (Kop-Flex, Inc.).

quate lubrication produces high sliding forces, meaning that any misalignment will produce high reactions on the coupled machines. Clearance between gear teeth, particularly as the coupling wears, allows the cover to take a slightly different radial position with each start, which poses a major balance problem in high speed equipment.

Horizontal motors with sleeve journal bearings generally do not have a thrust bearing to position the rotor. They rely, instead, on the pump's thrust bearing via a limited end float coupling. Figure 12.11 shows the principle. Flexible metal disc and diaphragm couplings inherently limit end float. All other flexible coupling types require special provisions to realize this capability.

Rigid couplings (Fig. 12.12) transmit torque by shear in fitted bolts, bending by tension in the same bolts. Most designs use a rabbet or spigot fit to maintain concentricity between the hubs.

Pumps arranged for dismantling without disturbing the driver, or for bearing and seal maintenance without opening the casing, require a spacer coupling. All-metal couplings are easily furnished with a spacer, the sole caution being for spring grid couplings which incorporate the coupling in the spacer, an arrangement that increases the overhung moment. Elastomer couplings are radially flexible, at least in terms of dynamic balance; thus one end of a spacer arrangement must be rigid. A poorly designed rigid connection or high misalignment forces or both can lead to catastrophic failure of the connection.

For a given parallel misalignment, the angular deflection of disc, diaphragm, and gear-type couplings decreases with increasing coupling element separation. Many users recognize this and specify a minimum spacer length to ensure adequate coupling element separation.

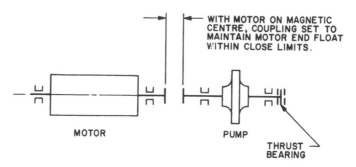

FIG. 12.11. Limited end float coupling arrangement.

In installations involving low speed pumps with the driver well removed from the pump, the use of universal drive shafts is justified. The application of such drive shafts (see Fig. 12.13) needs care. Critical matters are bearing life in the universal joints (Hookes couplings), shaft critical speeds, and the need for some misalignment to ensure sufficient movement of the joint bearings.

On occasion, pumps are arranged for dual drive, e.g., motor and turbine. Often it is considered desirable to not rotate both drivers all the time, so some

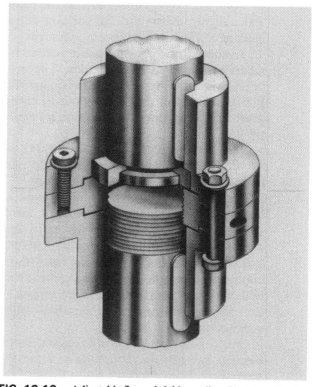

FIG. 12.12. Adjustable flanged rigid coupling (Kop-Flex, Inc.).

FIG. 12.13. Tubular intermediate universal joint shafting (Industrial Power Systems Division, Dana Corp.).

form of overrun becomes necessary. In the simpler arrangement only the standby driver has an overrun clutch. The clutch can be either centrifugal (rotation throws "leading" shoes into contact with a drum; see Fig. 12.14) or a Sprag type (cam-shaped elements arranged to lock in one direction; see Fig. 12.15).

The couplings described so far are all direct drives, i.e., the driver and driven shafts rotate at the same speed. For some applications it is desirable to retain the simplicity of fixed speed drive but have the pump run faster or slower than its driver. A V belt drive is the usual means of doing this. Figure 12.16 shows a typical arrangement. V belts transmit torque by tension, thus

FIG. 12.14. Centrifugal-type clutch coupling (Centric Clutch Division, Zurn Industries, Inc.).

FIG. 12.15. Sprags are kept in contact with members by energizing springs; wedge tight for one direction of drive and release for the other direction (Courtesy Dana Corporation, Formsprag, Warren, Michigan).

FIG. 12.16. V belt drive, motor mounted overhead (Worthington Pump, Dresser Industries, Inc.).

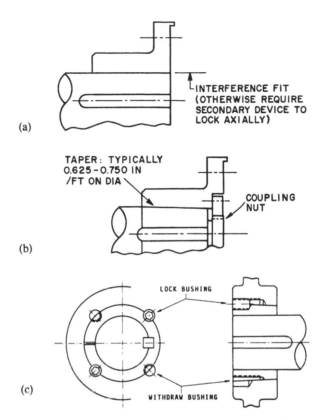

(a)

(b)

(c)

FIG. 12.17. Coupling hub mountings. (a) Cylindical fit. (b) Taper-mounted coupling. (c) Taper lock bushing.

imposing a substantial radial load on the connected machine shafts. Unless the shaft extensions and adjacent bearings are designed to accommodate this load, the power that can be transmitted is limited lest the shaft or bearings or both suffer premature failure. Using a jack shaft overcomes this limitation but complicates the drive.

For the types of pumps used in chemical processing, couplings are mounted one of two ways: keyed cylindrical fit or keyed tapered fit. The former is less expensive and is therefore more common. Cylindrical fits (Fig. 12.17a) are better if sized for interference. Slide fits seem attractive for ease of mounting and dismounting, but are prone to fretting, fatigue, and require a secondary device for axial location. With the appropriate tools, interference fits do not pose a major mounting and dismounting problem. In pumps designed for bearing and seal maintenance without opening the casing, a taper-mounted coupling (Fig. 12.17b) offers easy mounting and dismounting without the limitations of cylindrical slide fits. Taper mounting couplings increase the first cost of the pump. For low speed couplings the use of a taper lock bushing (Fig. 12.17c) offers an easily assembled interference fit mounting. The disadvantages are an increase in coupling size to accommodate the taper bushing and greater difficulty maintaining concentricity. The latter problem affects balance, hence limits speed.

References

12.1. M. Polk, "Better Couplings Reduce Pump Maintenance," *Hydrocarbon Process.*, pp. 62, 63 (January 1987).

12.2. C. B. Gibbons, "Coupling Misalignment Forces," in *Proc. 5th Turbomachinery Symp.*, October 1976, pp. 111–116.

12.3. J. R. Mancuso, "Disc vs Diaphragm Couplings," *Mach. Des.*, pp. 95–98 (July 24, 1986).

12.4. J. Wright, "Principal Engineering Features Required by High Performance Gear Couplings," in *Proc. Int. Conf. on Flexible Couplings, UK*, June 19–July 1, 1977, pp. B4-1 through B4-15.

13. Baseplates

In functional terms, a pump has to be connected to its driver in a manner that will maintain acceptable alignment under drive reaction, thermal expansion, and reasonable piping loads. Close coupling the pump and driver achieves this by direct connection, thus avoiding the need for earthbound support; see Anderson [6.26].

Close coupling is not always possible or agreeable. Relative size of pump and driver, pump rotor support requirements, and tradition often dictate that pump and driver remain separate machines coupled together. In its simplest form a separately coupled pump and driver can be bolted directly to a foundation, the foundation being stiff enough to satisfy the functional requirements of the pump-to-driver structural connection.

Equipment bolted directly to a foundation is not easily moved or shimmed to correct alignment. Adding soleplates, usually under the driver, overcomes this deficiency but complicates installation since the soleplates must be carefully positioned and leveled.

The installation difficulties associated with soleplates can be reduced, to varying extents, by mounting the equipment on a base which is in turn attached to the foundation (see Fig. 13.1). Along with simpler installation, mounting the equipment on a base simplifies shop assembly and unit shipping. For these three reasons most horizontal pumps are furnished with a base under the pump and driver.

Since the base is an element in the structural connection between pump and driver, its performance is crucial to satisfactory operation of the equipment. Despite its importance, base integrity is frequently sacrificed to price, often to the great detriment of the pump or driver or both. Murray [13.1] drew attention to this in 1973, and Nailen [13.2] saw fit to repeat the caution in 1988.

Materials for bases are cast iron, carbon steel, austenitic stainless steel, or reinforced polymers. Cast iron is limited to standard bases. Fabricated carbon

FIG. 13.1. Baseplate under pump and driver; bolted to the foundation.

steel allows design flexibility and is the usual material. When baseplate corrosion is a problem, austenitic stainless steel or reinforced polymer offer two solutions. Each of these materials has particular design considerations. Since fabricated steel is the most common material, the following discussion centers on steel with notes on the other materials where applicable.

Bases are divided into two types: supported and suspended. The distinction derives from how the base is connected to and supported by the foundation. Supported bases are bolted and grouted to a "stiff" foundation (see Fig. 13.2). Once the base is installed, its function is to transmit equipment loads and drive reaction to the foundation. The remainder of the structural connection between the equipment, bending and torsion, is then provided by the foundation. Noting how they function, supported bases are designed for:

Moderate bending stiffness; sufficient for handling as an assembled unit without yielding.
High pedestal and foundation connection stiffness; sufficient to maintain coupling alignment under imposed loads.

Torsional stiffness is not a design requirement for supported bases because the foundation provides it. This reduces cost but at the expense of installation cost, since the base must be leveled after positioning it on the foundation.

A side effect of the deliberate lack of torsional stiffness is that precise shop alignment of the mounted equipment is meaningless; the alignment will

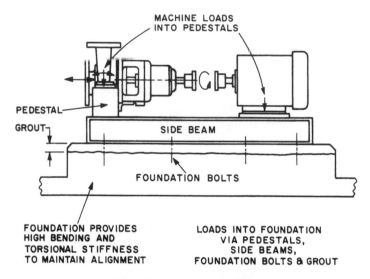

FIG. 13.2. "Stiff" foundation.

change when the unit is moved. What is important in shop alignment is that with the equipment mounting surfaces level, the alignment is "close enough" to allow precise alignment in the field. For this same reason, drivers should not be doweled until after final alignment in the field.

Suspended bases, for want of a better term, are those that do not rely on the foundation for torsional and bending stiffness. They are used in the following circumstances:

Simplified installation; the unit is prealigned and can be positioned, connected, and started. Any attachment to the foundation is nominal and usually three point to be self-leveling.

Minimize structure-borne noise; the unit is suspended above the foundation on resilient mounts (springs or elastomer).

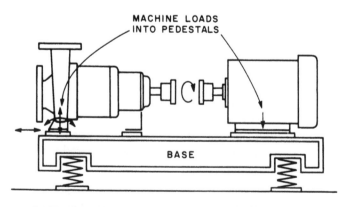

FIG. 13.3. Flexibly mounted baseplate.

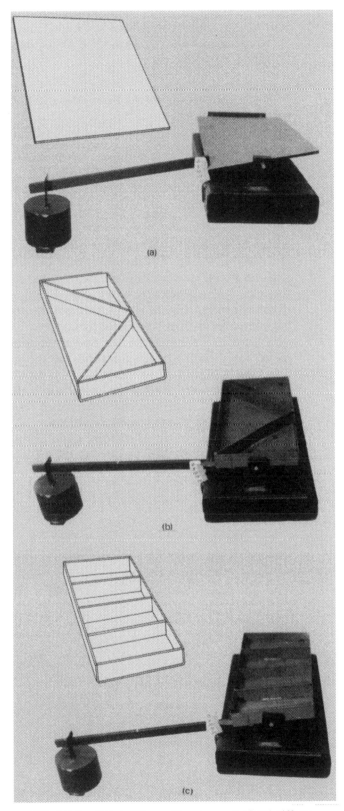

FIG. 13.4. Effect of diagonal bracing on torsional stiffness.

Reduce piping loads; the unit is suspended on springs or stilts or free to slide so it can move to accommodate piping expansion. Figure 13.3 shows a spring-mounted base.

Compared to supported bases, the additional design requirements for suspended bases are:

High bending stiffness; sufficient to maintain coupling alignment under equipment weight and imposed loads.
High torsional stiffness; sufficient to maintain coupling alignment under drive reaction.

These two additional requirements raise the base cost. Both requirements add weight, though not grossly if well designed, but providing torsional stiffness involves more difficult fabrication. As Blodgett [13.3] points out, torsional stiffness requires either closed sections (undesirable in bases) or diagonal bracing. Figure 13.4 illustrates quite graphically how diagonal bracing compares to cross bracing for torsion.

Figure 13.5 summarizes the design requirements for the two types of bases. Base bending design guidelines are shown in Fig. 13.6. For supported bases the values of W at the equipment mounting points need only take account of imposed loads if the base is stiffly attached to a foundation. This assumes that the stiffness characteristic of resilient supports matches predicted movements so the reactions are low.

Guidance on torsional stiffness requirements is given in Fig. 13.7. The torque applied to a base is usually just the drive reaction. Check, however, for

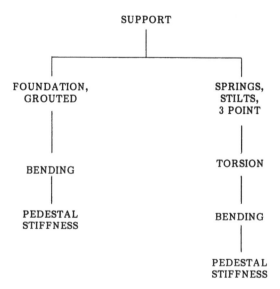

FIG. 13.5. Design requirements for rigidly (left) and flexibly (right) mounted bases.

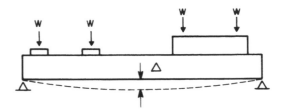

BASES

BENDING

$$\Delta \begin{cases} 0.00025 \ \text{IN/IN} - \text{GROUTED} \\ 0.0001 \ \text{IN/IN} - \text{UNGROUTED} \end{cases}$$

- CONSIDERATION AS SIMPLY SUPPORTED BEAM IS CONSERVATIVE

- Δ FOR GROUTED ALLOWS FOR SAFE HANDLING; RELIES UPON FOUNDATION FOR HIGH STIFFNESS

- Δ FOR UNGROUTED PROVIDES HIGH STIFFNESS IN BASE

FIG. 13.6. Guidelines for design in bending.

overhung weights and interaction between piping loads if the drive is a steam or liquid turbine.

Pedestals should be designed as high stiffness beams in their own right, well connected to the base, and not dependent upon grouting. Figure 13.8 details deflection criteria and illustrates "good" and "poor" design. Pedestals for two-point, centreline supported pumps pose a special problem. The limiting criterion is pump displacement at the coupling. Figures 13.9 and 13.10 offer some guidance on pedestal design. Side beam vertical deflection between the bolts can be a factor, too. Bussemaker [13.4] draws attention to the significant influence of pump stiffness under imposed loads. Tall pedestals are better avoided. Where they are unavoidable, their design must consider both deflection and resonance, the latter for bending and torsion.

LIMIT 'DYNAMIC' MISALIGNMENT TO ≃ 0.002″

DUE BASE TWISTING

MUST USE BOX SECTION OR DIAGONAL BRACING; LATTER PREFERRED. CROSS BRACING NOT EFFECTIVE IN TORSION

FIG. 13.7. Recommended torsional stiffness criterion for flexibly mounted bases.

- DO NOT RELY ON GROUT FOR SUPPORT, PEDESTAL MUST BE
  BEAM SUPPORTED @ SIDE BEAMS ONLY

- DEFLECTION: LESSER OF

    Δ = 0.0001 IN/IN

    Δ DETERMINED FROM ALLOWABLE SHAFT MOVEMENT @
        COUPLING; PRODUCT OF IMPOSED FORCES

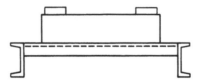

POOR - WEAK IN BENDING, SUPPORT BY GROUT CANNOT BE
      GUARANTEED

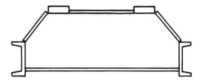

GOOD- PEDESTAL IS BEAM OF HIGH BENDING STIFFNESS

FIG. 13.8.   Pedestal design criteria.

Bases are designed for stiffness. A carbon steel base generally realizes this at minimum cost by a simple arrangement of moderately heavy pieces. In stainless steel, a more expensive material, lighter, more complex shapes are warranted to minimize cost. Cast iron is only half as stiff as carbon steel, so the sections need to be heavier, the increase sometimes being offset by the ease of producing more complex shapes. Reinforced polymers are an order of magnitude less stiff than steel, which limits their use in structures designed for stiffness. When they are used, the sections and configuration are necessarily quite different from those for metals.

Often the pump handles a liquid whose falling or accumulation on the foundation poses a problem (corrosion or combustion). The base is then required to also serve as a collector of incidental leakage. Two arrangements are in use; drip pan and drain rim. Figure 13.11 shows the essential difference. Drip pan bases offer an extensive sloping drainage surface but require careful design to ensure structural integrity and are difficult to fabricate. Drain rim bases are usually easier to design and fabricate. Their disadvantage is that the regions onto which pump leakage falls are generally flat, and thus they will accumulate some leakage. Openings in drip pans or the deck of drain rim bases have to be collared or bossed to avoid leakage through the opening.

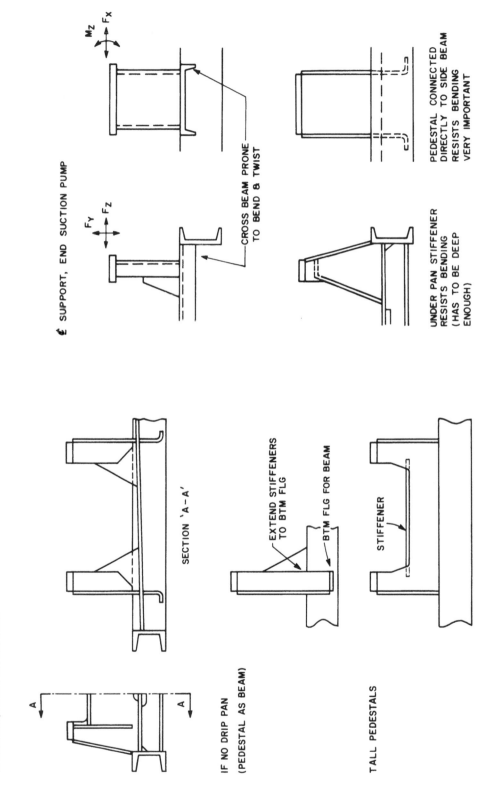

FIG. 13.9. Centerline support pedestal configurations.

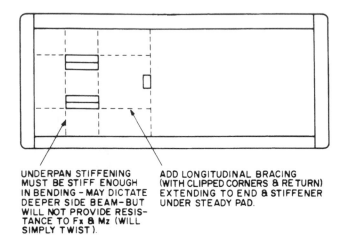

BASE UNDER PUMP WITH MANY AUXILIARIES; DIMENSIONS TO AVOID OVERHANG PRECLUDE ATTACHING PEDESTALS DIRECTLY TO SIDE-BEAMS.

UNDERPAN STIFFENING MUST BE STIFF ENOUGH IN BENDING - MAY DICTATE DEEPER SIDE BEAM-BUT WILL NOT PROVIDE RESISTANCE TO Fx & Mz (WILL SIMPLY TWIST).

ADD LONGITUDINAL BRACING (WITH CLIPPED CORNERS & RETURN) EXTENDING TO END & STIFFENER UNDER STEADY PAD.

FIG. 13.10. Pedestal support when direct connection to side beams is not feasible.

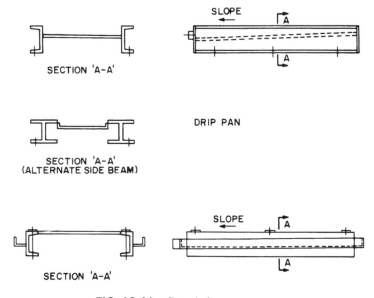

SECTION 'A-A'

SLOPE

DRIP PAN

SECTION 'A-A' (ALTERNATE SIDE BEAM)

SLOPE

SECTION 'A-A'

FIG. 13.11. Base drain arrangements.

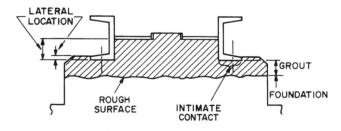

LATERAL LOCATION

GROUT

FOUNDATION

ROUGH SURFACE

INTIMATE CONTACT

FIG. 13.12. Functions of grouting.

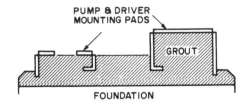

FIG. 13.13.   Grout-filled base.

Of necessity, supported bases must provide for grouting. If the base is well designed (see pedestals discussion above), the essential functions of grouting are to:

Ensure intimate contact between the base underside and the foundation.
Provide additional lateral restraint.

Figure 13.12 illustrates these.

A secondary function is to fill voids in or under the base to prevent the accumulation of liquid or debris or both. At one extreme, the base is an open structure designed to be filled with grout (Fig. 13.13); at the other, a closed structure designed for grouting to the underside of a drip pan (Fig. 13.14). Designs with drip pans or deck plates require special features to ensure grout can completely fill the void beneath the plate or pan. If the void is not completely filled, there is a risk liquid will accumulate under the plate, or the plate will "drum" and generate unnecessary noise. Figure 13.15 shows the features necessary for grouting under a drip pan.

Except for very small units, under say 500 lb, the base should include provisions for lifting. The lift is easier if it is four point. More points are feasible, but the rigging to achieve load sharing becomes complicated. The bending stiffness criterion given in Fig. 13.6 is good for four-point lift with the points at 1/4 span. Figure 13.16 provides additional guidance.

Because bases are designed for stiffness, the volume of welding required is not high. Weld extent and size combine to give an actual weld volume greater than required for stiffness. Continuous welding is necessary for all joints except for those under pan/plate stiffeners. Intermittent welding is not allowed because it is prone to corrosion. Weld size is that necessary to develop 50% of the plate strength.

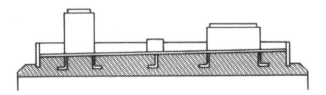

FIG. 13.14.   Grouted drip pan base.

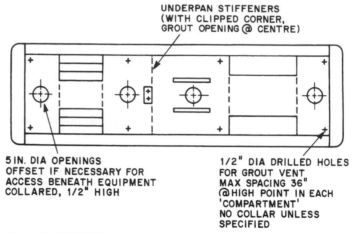

UNDERPAN STIFFENERS
(WITH CLIPPED CORNER,
GROUT OPENING @ CENTRE)

5 IN. DIA OPENINGS
OFFSET IF NECESSARY FOR
ACCESS BENEATH EQUIPMENT
COLLARED, 1/2" HIGH

1/2" DIA DRILLED HOLES
FOR GROUT VENT
MAX SPACING 36"
@ HIGH POINT IN EACH
'COMPARTMENT'
NO COLLAR UNLESS
SPECIFIED

API-610 REQUIRES
AT LEAST 1·19 SQ. IN.
OPENING IN EACH
UNDERPAN 'COMPARTMENT.'
NOT ALWAYS PRACTICABLE
eg. OPENING BENEATH EQUIPMENT

FIG. 13.15.   Design features necessary for grouting under a drip pan.

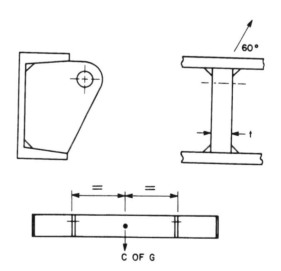

60°

t

C OF G

- 4 POINT LIFT (6 & 8 HARD TO EQUALIZE)

- EQUISPACED AROUND C OF G

- WATCH FOR SLING INTERFERENCE WITH EQUIP. PROTRUSIONS
  (JUNCTION BOXES, ETC.)

- 't' ≥ 'D' PIN DIA. – CHECK BENDING WITH 60° SLING
  INCLINATION

FIG. 13.16.   Lifting lug design.

STRESS RELIEVING
- IF REQ'D BY CUST SPEC'N OR ENGRG
- PREFFERED FOR BASES WITH EXTENSIVE
  FABRICATED PEDESTALS

MACHINING
- BASE SHOULD NOT BE 'SPRUNG' ON MACHINE TOOL BED
- MACHINE FOR 0.12 SHIMS UNDER DRIVER COMPONENTS
- MOUNTING SURFACES IN SAME PLANE WITH-IN
  0.002 PER FT OF SEPARATION

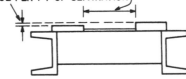

FIG. 13.17.   Requirements for base machining.

Either type of base is intended to be a precision surface on which to mount and accurately align the equipment. This function cannot be realized with an as-fabricated base, thus machining is necessary.

Bases with extensive welding should be stress relieved (preferably oven) before machining. Machining must allow for at least 0.12 shims under the driver components. Once machined, individual equipment mounting surfaces should be coplaner within 0.002 in. per foot of separation (see Fig. 13.17). Supported bases must be leveled by shimming at the foundation bolt holes before checking mounting surfaces for out-of-plane problems.

Additional refinements, mandated by some specifications, e.g., API-610, are leveling screws to aid base installation and equipment jacking screws for easier alignment.

References

13.1.   M. G. Murray, "How to Specify Better Pump Baseplates," *Hydrocarbon Process.*, pp. 68–71 (September 1973).

13.2.   R. L. Nailen, "Installing Motors Properly," *Plant Serv.*, pp. 72–76, 78, 80, 82 (January 1988).

13.3.   O. W. Blodgett, *Design of Weldments*, The James F. Lincoln Welding Foundation, May 1963, pp. 3.6–12.

13.4.   E. J. Bussemaker, "Design Aspects of Baseplates for Oil and Petrochemical Industry Pumps," *Proc. ImechE*, pp. 135–141.

14. Procurement

Successfully getting what's required, procurement, involves three steps:

1. Specification and bid request.
2. Evaluation and selection.
3. Order placement and expediting.

Broadly, procurement is the same for most items or articles, varying only in degree depending upon the complexity and cost involved. The following material therefore restricts itself to those matters either important to or peculiar to the procurement of pumps.

14.1. Specification and Bid Request

The intent of a specification is to state clearly and explicitly what is required and the desired terms of purchase. For ease of compilation, a specification is usually comprised of two sections: technical and commercial.

The technical section needs to address the points listed in Table 14.1. These points are must easily covered and least likely to be overlooked, if set out on a data sheet. If possible, the data sheet should be just that, a single page,

TABLE 14.1   Items to be Specified

Item	Reference
Plant, equipment number	Process sheets
Service, duty	Process sheets
Site conditions	Plant data
Extent of supply	Company practice
Liquid	Chapter 2
Conditions of service	Chapter 4
Pump:	
Hydraulic requirements	Chapters 5, 6, 7, 8, 9, 10
Configuration	Chapters 6, 7, 8, 9, 10, 11
Materials	Chapters 3, 6, 7, 8, 9, 10
Piping	Chapter 11 (for seal)
Driver	Reference 1.1
Drive arrangement	Chapters 12, 13
Environmental requirements	Prevailing regulations
Documentation	Company engineering standard
Inspection and tests	Company or industry standard
Protection	Site location, storage

FORM A1 CENTRIFUGAL PUMP DATA SHEET

PUMP SIZE AND MODEL _____ BRG. FRAME _____ SERVICE _____

NO. PUMPS REQ'D. _____ NO. MOTORS REQ'D. _____ ITEM NO. _____ NO. TURBINES REQ'D. _____ ITEM NO. _____

OPERATING CONDITIONS – EACH PUMP	PERFORMANCE

OPERATING CONDITIONS – EACH PUMP

LIQUID/SLURRY _____

PT. °F NORM _____ MAX. _____ US GPM AT NORM _____ RATED _____

SP. GR. at NORM PT. _____ TOTAL HEAD, FT RATED _____

VAP. PRESS. at NORM PT. PSIA _____ SUCT. PRESS. PSIG MAX. _____ RATED _____

VIS. at NORM PT SSu _____ NPSHA, FT _____

CORR./EROS. CAUSED BY _____ pH _____ HYD. HP _____

DRIVER HP TO BE SELECTED FOR MAX. S.G. _____ & MAX. VISCOSITY _____

CONSTRUCTION – ☐ ANSI B73.1 ☐ ANSI B73.2 ☐ OTHER _____

PUMP TYPE: ☐ HORIZ. ☐ VERT. IN-LINE ☐ COUPLED MOTOR SHAFT ☐ CRADLED MNT.

CASE HORIZONTAL MOUNT: ☐ FOOT ☐ CENTER LINE

  VERTICAL MOUNT: ☐ MOTOR SHAFT ☐ RIGID COUPLING ☐ OTHER _____

  SPLIT: ☐ RADIAL ☐ AXIAL TYPE VOLUTE: ☐ SINGLE ☐ DOUBLE

  PRESS: ☐ MAX. ALLOW. _____ PSIG _____°F ☐ HYDRO TEST _____ PSIG

  CONNECT: ☐ DRAIN ☐ GAGE SUCTION ☐ GAGE DISCHARGE

IMPELLER DIA. RATED _____ MAX. _____ IMPELLER TYPE _____

BEARINGS TYPE: RADIAL _____ THRUST _____

  LUBE: ☐ OIL ☐ OIL MIST ☐ GREASE ☐ GREASE FOR LIFE

COUPLING: MFR. _____ MODEL _____ GUARD _____ OILER _____

  DRIVER HALF MTD. BY. ☐ PUMP MFR. ☐ DRIVER MFR. ☐ PURCHASER

STUFFING BOX COVER: ☐ STANDARD ☐ JACKETED ☐ SEAL ONLY

PACKING: ☐ MFR. & TYPE _____ SIZE/NO. OF RINGS _____

  LANTERN RINGS: ☐ YES ☐ NO

MECH. SEAL: ☐ MFR. & MODEL _____ MATERIAL CODE _____

  ☐ BALANCED ☐ UNBALANCED ☐ SINGLE ☐ INSIDE ☐ OUTSIDE

  ☐ DOUBLE ☐ BACK TO BACK ☐ TANDEM ☐ FACE TO FACE

AUXILIARY PIPING (SEE FIG. NO. _____ FOR CODE)

☐ STUF. BOX PLAN NO. _____ ☐ C. W. PIPING PLAN NO. _____

TOTAL COOLING WATER REQ'D., GPM _____ ☐ SIGHT F.I. REQ'D.

☐ PACKING COOLING INJECTION REQ'D., TOTAL GPM _____ PSIG _____

  EXTERNAL SEAL FLUSH FLUID _____ GPM _____ PSIG _____

  SEAL QUENCH PLAN _____ SEAL QUENCH FLUID _____

DRIVER: ☐ MOTOR ☐ TURBINE ☐ OTHER PROVIDED BY _____

HP _____ RPM _____ FRAME _____ VOLTS/PHASE/HERTZ _____

MFR. _____ BEARINGS _____ SERVICE FACTOR _____

TYPE _____ INSULATION _____ AMPS: FL _____ LR _____

LUBE _____ TEMP. RISE °C _____ ENCL _____

INLET PRESSURE _____ EXHAUST PRESS. _____ STEAM TEMP. _____ WATER RATE _____

OTHER

PERFORMANCE

PERFORMANCE CURVE NO. _____

RPM _____ NPSH (WATER) _____

EFF. _____ % BHP RATED _____

MAX. BHP RATED IMPELLER _____

MAX. HEAD RATED _____

MAX. DISCH. PRESS. PSIG _____

MIN. CONTINUOUS GPM _____

SHOP TESTS

☐ NONWIT. PERF. ☐ WIT. PERF.

☐ NONWIT. HYDRO. ☐ WIT. HYDRO.

☐ NONWIT. NPSH ☐ WIT. NPSH

☐ NONWIT. VIBRATION ☐ WIT. VIBR.

☐ DISMANTLE & INSPECT AFTER TEST

☐ OTHER: _____

PUMP MATERIALS

CASING _____

IMPELLER _____

WEAR RINGS _____

SHAFT/SLEEVE _____

GLAND _____

GASKETS _____

BASE PLATE _____

COUPLING GUARD _____

OTHER: _____

INSPECTION ☐ NOT REQUIRED

  ☐ IN PROCESS ☐ FINAL

  _ DAYS NOTIFICATION REQUIRED

SOUND SPECIFICATION REQUIREMENTS

ADDITIONAL REQUIREMENTS/COMMENTS

CUSTOMER/ USER _____

LOCATION _____

CUSTOMER P.O. NO. _____

ITEM NO(S). _____ EQUIP. NO(S). _____

FACTORY ORDER NO(S). _____ PUMP SERIAL NO(S). _____

ISSUED BY _____ DATE _____

REV. _____ DATE _____

FIG. 14.1. Centrifugal pump data sheet.

otherwise it becomes cumbersome to use and tends to be neglected. Several of the prevailing industry standards include a recommended data sheet, although they are not always one page. Figures 14.1, 14.2, and 14.3 show typical simple data sheets for centrifugal, rotary, and reciprocating pumps, respectively.

The data sheet should anticipate the evaluation process by including those items needed for evaluation. Included in these are pressure ratings, load ratings, bearing sizes, and other dimensions or coefficients to enable a comparison of the equipment offered.

Ideal specifications are functional, i.e., they state what has to be done, such as performance and life, but do not place any restrictions on how it is to be done. For a variety of reasons, some valid, some not, such specifications are very much in the minority. The usual industry specification tends to reflect the experience of that industry and places corresponding limitations on how the equipment is to be constructed. The virtue of this is that it eases the task of evaluation and ensures relatively uniform equipment. It does not, however, preclude the need for common sense in selecting equipment. And provided common sense is applied, there is every reason to encourage manufacturers to offer alternative designs or configurations; there is usually more than one way to achieve a particular result.

Commercial conditions that should be addressed in the specification include:

Delivery required
Terms of payment and escalation of applicable
Performance guarantees
Warranty required
Penalties and reciprocal bonuses
Default provisions

When the bid request consists of a number of documents, some peculiar to the application, some general, it is important to at least state the hierarchy of those documents. Usually this is: requisition or order, data sheet, general company specifications, general industry specifications.

Citing prevailing government regulations needs special care. To say that the equipment must satisfy "any and all governmental regulations" is a convenient catchall but does very little to ensure the equipment complies. To aid compliance, the relevant regulations should be cited and the current or cut-off date given.

Of the two approaches available for getting bids, public tender and prequalified bidders, the latter generally yields better results. It does so because the vendor is selected from a number whose competence has been previously established by demonstration and assessment. Vendor assessment is as important in vendor prequalification as past experience, particularly if past experience is more than a year old. Two factors account for this. First, a vendor's organization is subject to change. Second, it is possible to produce very good equipment with a totally people-dependent system. Quality real-

DRIVER DATA

MANUFACTURER _____ HP _____ RPM _____ PHASE _____

CYCLES _____ VOLTAGE _____ INSUL. _____ FRAME _____ S.F. _____

ODP _____ TEFC _____ EXP. PROOF _____ OTHER _____

RELIEF VALVE (IF REQUIRED)

MANUFACTURER _____ MODEL _____ SIZE _____ SET PRESS. (PSIG) _____

SPEED REDUCER (IF REQUIRED)

MANUFACTURER _____ MODEL _____ RATIO _____

DRAWINGS

	APPROVAL				FINAL								
	OUTLINE	DRIVER	CROSS SECTION	OTHER	OUTLINE	DRIVER	CROSS SECTION	SPARE PARTS	INSTR MAN	PERF CURVES	MECH SEAL	OTHER
NORM. QTY.	0	0	0		3	0	0	3	3	0	0	
NO. REQUIRED												
NO. SENT												
DATE SENT												

INSPECTION AND TESTING

	WITNESSED		
	YES	NO	
FINAL INSPECTION		YES	NO
CERTIFIED PERFORMANCE TEST			
CERTIFIED NPSH TEST			
CERTIFIED HYDRO TEST			
OTHER			

TO BE COMPLETED BY ENGINEERING

TEST STATION # _____

_____ H.P. MOTOR

_____ IN. LB. TORQUE METER

SPECIAL NOTES - EXCEPTIONS TO SPECIFICATIONS, ETC.

WC322

DRESSER PUMP DIVISION (DRESSER)

SIER BATH ROTARY PUMP DATA SHEET

CONTRACT NO. _____

S.O. NO./INQ. NO. _____

SALES OFFICE/REP. _____

DATE: _____

CUSTOMER: _____

CUSTOMER ORDER NO. _____

CUSTOMER TAG NO. _____

WRITTEN BY: _____

PAGE _____ OF _____

ADDITIONAL NO. _____

ITEM NO. _____

NO. OF UNITS. _____

CONTRACT TO SPEC. _____

SPEC. NO. _____

SERIAL NO. _____

CONDITIONS OF SERVICE

SERVICE _____

FLUID _____

CAPACITY _____ (USGPM) _____ (M³/Hr.)

DISCH. PRESS _____ (PSIG) _____ (Kg/Cm²g)

SUCT. PRESS _____ (PSIG) _____ (Kg/Cm²g)

DIFF. PRESS _____ (PSI) _____ (Kg/Cm²g)

REL. VALVE SET _____ (PSIG) _____ (Kg/Cm²g)

VISCOSITY _____ (SSU) _____ (CP)

SP. GRAVITY _____ @ P.T.

PUMPING TEMP. _____ (°F.) _____ (° C.)

NPSHA _____ (Feet) _____ (Meters)

VAPOUR PRESS. _____ (PSIA) _____ (Kg/Cm²a)

QUOTED PERFORMANCE

CAPACITY (USGPM) _____ (M³/Hr)

PUMP RPM _____

BHP (DESIGN) _____ (Kw)

BHP (R.V. Setting) _____ (Kw)

NPSH REQ'D. (FEET) _____ (Meters)

PUMP DESCRIPTION

PUMP SIZE _____ MAT'L OF CONST. _____

BEARINGS - INTERNAL ( ) EXTERNAL ( )

ROTOR WIDTH/PITCH _____

FLOW-STANDARD ( ) REVERSE ( )

CLEARANCES _____

SUCTION CONNECTION _____

DISCHARGE CONNECTION _____

SUCTION ORIENTATION _____

DISCHARGE ORIENTATION _____

DRIVE SHAFT ORIENTATION _____

ROTATION _____ ROTOR SET-UP _____

SCREWS - PINNED ( ) INTEGRAL ( )

SCREW TIPS - BRONZE ( ) HARD TIPPED ( )

SEALING - PACKED ( ) MECHANICAL ( )

SEAL TYPE - _____

BASEPLATE - STD. ( ) REDUCER ( )

NAVY PEDESTAL ( )

COUPLING ( ) TYPE - _____

COUPLING GUARD ( )

OPTIONS (✓) IF REQUIRED

SPLIT PACKING GLANDS ( )

LANTERN RINGS - STANDARD ( ) SPLIT ( )

HIGH TEMPERATURE CLEARANCES ( )

HIGH TEMPERATURE OIL SEALS ( )

HIGH TEMPERATURE PACKING ( )

OIL COOLING COILS ( )

OTHER: _____

HOPPER BODY ( )

JACKETED BODY ( )

JACKETED STUFFING BOXES ( )

CHROMED BORES - 001" ( ) .005" ( )

- OTHER: _____

EXTENDED SCREWS ( )

CENTER BEARINGS ( )

HARDENED SHAFTS AT PACKING ( )

FIG. 14.2. Typical rotary pump data sheet (Worthington Pump, Dresser Industries, Inc.).

261

WORTHINGTON PUMP CORPORATION
ENGINEERED PUMP DIVISION

POWER PUMP DATA SHEET

ORDER NO. _____
DATE _____
QUOTE NO. _____

PUMP SIZE AND TYPE _____

REVISIONS	NO.	DATE	DESCRIPTION	NO.	DATE	DESCRIPTION	NO.	DATE	DESCRIPTION

OPERATING CONDITIONS

LIQUID PUMPED _____
PUMPING TEMP. _____ °F.
SPEC. GR. @ P.T. _____
VISC @ P.T. _____ CP/SSU
V.P. @ P.T. _____ PSIA
CORR/EROS CAUSED BY

CAPACITY @ P.T. _____ GPM
DISCHARGE _____ PSIG
SUCTION _____ PSIG
DIFFERENTIAL _____ PSI
NPSH AVAILABLE _____ FT.
HYDRAULIC HP

ENVIRONMENT

AMBIENT TEMP.   AV _____ MIN. _____ MAX. _____
INSTALLATION   ☐ INDOORS   ☐ OUTDOORS
               ☐ HEATED   ☐ WITH ROOF
               ☐ UNHEATED   ☐ WITHOUT ROOF
ELEVATION _____
COOLING WATER _____ °F.   PRESSURE   FRESH   SEA

PERFORMANCE		TECHNICAL DETAILS		MATERIALS	PUMP OPTIONS
NPSHR FT.	NPSHA FT.	PLUNGER DIA.		☐ STANDARD ☐ OTHER	YES OPTIONS
RPM AT DES. GPM	RPM MAX.	STROKE		CYLINDER	☐ SYNCRONIZED SUCTION VALVE UNLOADERS
PLUNGER SPEED FT/MIN @ DES. GPM		NO. OF PLUNGERS		PLUNGER	☐ SUCTION PULSATION DAMPER
MECH EFFICIENCY %		VALVES – SUCTION		STUFFING BOX	☐ DISCHARGE PULSATION DAMP
VOL. EFFICIENCY		AREA VALVE		THROAT BUSH.	☐ OIL COOLER
HYDRO TEST PRESSURE		VALVE VELOCITY DES.	MAX.	PACKING GLAND	☐ AUTO BY-PASS CONTROL VAL
FRAME LOAD MAX. OPER. R.V.		VALVES – DISCHARGE		LANTERN RING	☐ INDIVIDUAL VALVE ACESS
FRAME LOAD AT DES. PRESS OPER. PRESS		AREA/VALVE		VALVE	☐ SEGMENTED LIQUID CYLINDER
		VALVE VELOCITY DES.	MAX.	VALVE SEAT	☐ JACKETED STUFFING BOXES
				VALVE SPRINGS	☐ STEAM JACKETED CYLINDERS
				SPRING RETAINER	☐ COOLED DISTANCE PIECE
				SIDE RODS	☐ PACKING LUBRICATOR
				PACKING	☐ TACHOMETER
				OTHER – SPECIFY	☐ POWER TAKE-OFF
					☐ AUX. LUBE SYS.
					☐ OTHER – SPECIFY

LUBRICATION AND COOLING		PACKING	MAIN	AUX.		RELIEF VALVE
FRAME LUBE ☐ SPLASH ☐ PRESSURE		MANUFACTUER				MFG _____ TYPE _____
LUBE PIPING BY:						SET PRESSURE
PACKING LUBE ☐ OIL ☐ GREASE ☐ FLUSH		NO. OF RINGS				ACCUMULATION
FLUSH SOURCE _____ GPM		TYPE				
LUBRICATOR MAKE _____						
SIZE NO. FEEDS		PACKING SIZE				
COOLING ☐ PLUNGER ☐ STUFF BOX ☐ LIQ. CYL.		FLANGES	SIZE/POS	RATING	STUFFING BOX SEALING/FLUSHING MEDIUM	
C.W. _____ °F PSIG GPM		SUCT			HEATING	
C.W. PIPING BY :		DISCH				

ACCESSORIES		MOTOR	AUX PUMP	STEAM TURB	ENGINE	GEAR	SPEED VARIATOR
MFG							
TYPE							
DESIGN							
SPEC NO.							
DATA SHEET							
HP/RPM						RATIO	SPEED VAR.
OTHER		PH V. CY.				SERVICE FACTOR	
WEIGHT: PUMP BASE MOTOR TURBINE ENGINE GEAR							

INDUSTRY STANDARDS _____
CUSTOMER SPECIFICATION(S) NO.(S) _____

WM-184

FIG. 14.3. Typical reciprocating pump data sheet (Worthington Pump, Dresser Industries, Inc.).

ized this way can change markedly with personnel. The crucial aspect in vendor assessment is to go to the plant where the product will be made and talk to the people who will produce it.

14.2. Evaluation and Selection

Practically all pump purchases warrant some degree of evaluation. The usual procedure is to set down the data to be compared in tabular form in a so-called "bid tabulation." Data sheet format can aid this process by direct comparison or development of a tabulation by "cut and paste."

Most evaluation procedures involve two steps: assessment of mechanical equivalence, during which some offers may be eliminated, then an assessment of the "cost of ownership." In general terms, "assessing mechanical equivalence" requires the comparison of

1. Pressure containment, casing stiffness.
2. Bearing bracket or power frame integrity, i.e., shaft stress and stiffness, bearing capacity, and margin over loading.
3. Materials of construction.
4. Alternative configurations if offered.

Determining the "cost of ownership" involves the development of:

1. An estimate of installation cost, taking account of items such as motive power supply, NPSHR, piping difficulty, bypass size, base leveling, and equipment alignment.
2. Energy consumption evaluated over the expected operating profile. Efficiency comparison for a single point is not meaningful; the evaluation should make due allowance for accuracy, typically no better than 2% when power has been calculated from rating curve efficiency. Note, too, that in chemical process pumping the cost of pumping power is relatively low compared to the total energy consumed and the cost of lost production, therefore the ranking of energy consumption should not assume the same importance as it does in, say, pipelining.
3. A weighting factor to account for predicted maintenance cost. Quite a deal of judgment is necessary to do this, and quite a deal of courage is often necessary to defend it, but doing so is necessary to avoid the purchase of equipment that can cost far more to own than its low purchase price saved. Comparisons developed for mechanical equivalence are useful for this. Note for centrifugal pumps that low NPSHR may lower installation cost while increasing maintenance cost; see Chapter 6.

Buehler [14.1], Neerken [14.2], and Heisler [in Ref. 1.1] provide further detailed comment on pump evaluation and selection.

Summing the purchase price and the evaluation factors detailed above produces the evaluated cost. Equipment selection is generally based on this. Factors that can result in a contrary decision are: existing population (often factored into maintenance cost but shouldn't be), quoted delivery, or unacceptable commercial conditions. The last is a matter beyond the scope of this text.

A factor of major importance, particularly when delivery is critical, is to make the equipment selection promptly and get it on order so manufacture can start. When procrastination sets in, the result is an attempt to produce the equipment in less time to keep the overall program on schedule. Such haste often results in poor quality equipment.

14.3. Order Placement and Expediting

Beyond the caution sounded above on the need to decide and order promptly, the mechanics of placing the purchase order are outside the scope of this writing. The need for promptness, however, cannot be overemphasized; modern purchasing practice seeking to cover all the aspects can sometimes lose sight of the need for promptness in placing the order.

Expediting should not be left until the pump is nearly due. The better approach is to follow up the vendor, at the point of manufacture, shortly after order placement, to make sure the order is in production. As with purchasing procedures, order processing procedures can go awry and fail to get the order started in time. There is an adage in the engineering/construction industry that says "An order goes wrong in the first four weeks." For equipment that is not off the shelf, the adage carries a great deal of truth.

References

14.1. M. W. Buehler, "Centrifugal Pump Evaluation: A Systematic Approach," *Chem. Process.*, *49*(1), 86–94 (1986).
14.2. R. F. Neerken, "Progress in Pumps," *Chem. Eng.*, pp. 76–88 (September 14, 1987).

15. Installation

The successful installation of any pump has two phases: design and physically putting the equipment in place. Design is better if addressed before the specification is written. Not doing this increases the risk that a major requirement will be overlooked and inadequate equipment bought. When the pump is large, the means of physically installing it also need consideration before writing the specification, since shipping and handling may be a problem.

Noise is an increasingly important consideration in pump installations. Pump selection, installation, and operation all have a bearing on the end result. The topic is specialized; see Brennan [15.1], Szensai et al, [in Ref. 1.1] and Harris [15.2].

15.1. Location

In determining where to place the pump, the following points should be considered:

Pump as close as possible to its suction source.

Liquid level, or at least the high liquid level, above the pump to avoid the need for priming.

Freedom from windblown dust or sand, water sprays (from washing down operations), and flooding.

Access for routine inspection.

Room around, lifting provisions, and access to for dismantling the pump and its driver; see the manufacturer's outline drawing.

Risk of passing traffic accidently damaging the unit.

15.2. Support

Most horizontal pumps are supported on a "massive" foundation. The general rule of thumb for sizing such foundations is three times the unit mass for centrifugal and rotary pumps, five times for reciprocating pumps. As well as providing mass, the foundation must be designed to provide the bending and torsional stiffness necessary to maintain alignment between the pump and its driver. To avoid the influence of adjacent floor movement, "massive" foundations are generally cast separately from the floor. Equipment supported by a "massive" foundation is of necessity grouted to the foundation. The usual practice is to allow at least 2 inches between the base flange and foundation top for grout. See Fig. 13.2 for a typical "massive" foundation.

When a pump is to be supported by a structure, either the structure or the pump base must be designed to provide the bending and torsional stiffness necessary to maintain equipment alignment. Chapter 13 covers the design of such bases. In the case of reciprocating pumps, the structure also has to accommodate the pump's shaking forces.

Resiliently mounted pumps require special bases (see Chapter 13), but the foundation or structural support design has only to consider equipment

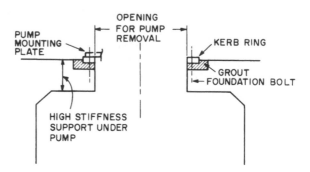

FIG. 15.1.   Typical foundation for vertical wet pit pump.

weight plus whatever forces piping displacement produces. Three-point supported pumps, relatively rare at present, have the same base requirements as resiliently mounted pumps, but the foundation will be subjected to the full connected piping loads since the connection to the foundation has high vertical and lateral stiffness.

Vertical wet pit pumps are generally supported on a reinforced concrete structure. The stiffness of this structure has a profound influence on the natural frequencies of the installed pump. The design therefore needs care to ensure the stiffness is not so low that it brings a natural frequency into close proximity of the running speed, thus causing a resonant vibration problem. Figure 15.1 shows a typical vertical wet pit pump foundation and the "kerb ring" generally employed to enable ready removal of the pump.

15.3. Suction System

Some 60% or so of pump problems originate in the suction system; therefore a good design goes a long way toward assuring a sound installation. By way of a summary, the points to be considered in suction system design are:

Vessel or sump configuration
Exit from the vessel or sump
Piping length, size, configuration
Priming, if required, and venting
Need for a temporary strainer

The suction vessel or sump must be designed to avoid air entrainment and uneven flow distribution. Air can be entrained by "above the liquid" flows into the vessel or by vortexing produced by the confluence of flows. Feed to the vessel and any liquid returns should therefore be below the storage liquid level and separated from the outlet by a weir, the latter to avoid direct communication. Sumps or forebays have to be designed for low velocities and

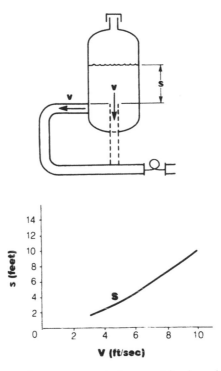

FIG. 15.2. Minimum submergence required over a plain pipe exit to prevent vortexing.

arranged to minimize flow disturbances. The Hydraulic Institute Standards [5.1] provide conservative guidance for such designs. Less conservative designs are feasible but should not be used unless verified by model testing.

Unless the exit from the vessel or sump has sufficient submergence (see Fig. 15.2), the outgoing flow will produce vortices with consequent air or gas entrainment. Depending upon the degree of gas entrainment, the pump may suffer from reduced performance, surging, or cessation of flow. Figure 15.2 shows the minimum submergence needed with a plain pipe exit. By using a "vortex breaker" at the exit, the submergence needed can be reduced. Figure 15.3, a drawing from Kern [in Ref. 1.2], shows various forms of these devices. In employing vortex breakers, it is important to appreciate that they do not provide immunity to vortexing, just reduce the submergence needed to prevent it. Data on how much the submergence can be reduced are lacking and need to be inferred from similar arrangements or determined by testing. When suction system head loss is critical, a bell-mouth or rounded exit will help reduce the loss. Vertical wet pit pumps take their suction directly from a sump. Location of the pump suction relative to the sump and suction submergence are important. As with sump dimensions, the Hydraulic Institute Standards provide conservative guidance.

Suction piping should be kept as short as possible. Three factors justify this requirement: head loss, inertia, and residence time. The first factor is self-explanatory. Inertia has to do with the risk of separation in the line should the pump be able to accelerate the discharge mass faster than the suction source

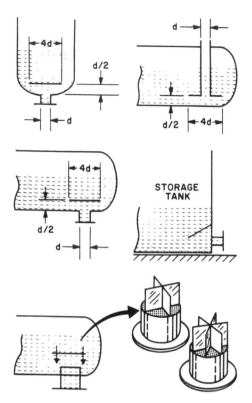

FIG. 15.3. Various "vortex breaker" arrangements.

can accelerate that of the suction. Inertia is an occasional factor in centrifugal and rotary pump installations, but for reciprocating pumps it is critical; see Chapter 8. Residence time is important in processes involving a risk of quenching a saturated suction vessel during an operating transient. Open cycle power plants are one example; see Karassik [15.3] for a detailed discussion.

The sizing of suction piping has to take account of pump suction nozzle size, suction system head loss, and residence time. Suction piping must be at least as large as the pump's suction nozzle, and is usually one size larger. For a manifold to multiple pumps, the area at any section should equal the sum of those areas downstream. Head loss should always be calculated, and the line size increased if necessary to ensure adequate NPSHA. Residence time requires that the suction line not be unnecessarily large lest the residence time be needlessly increased.

Qualified by the need to minimize piping loads on the pump (see below) the suction piping configuration should be as simple as possible. The run should slope in one direction, without high points that can accumulate gas or vapor. With the liquid below the pump, a suction lift, the slope is up toward the pump. When the liquid is above, a "flooded" suction, the slope is up toward the suction vessel. Elbows incur additional head loss and, perhaps more importantly, produce flow distortions, so should be used only when necessary. The flow distortions produced by elbows preclude their use in a

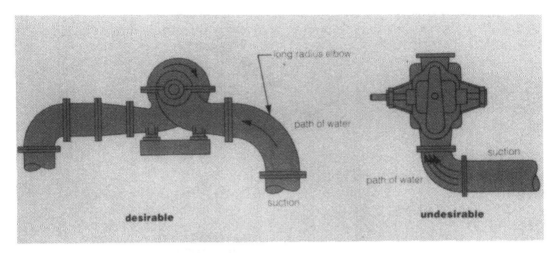

(a)

(b)

FIG. 15.4. (a) Elbow at the inlet of a double suction pump must be installed in a plane at right angles to the pump shaft, as shown at left. Orientation shown at right may cause serious operating problems. (b) Preferred arrangement of reducer and long radius elbow on a single suction pump when straight pipe approach cannot be provided. Note: If suction source is below the pump, the reducer should be eccentric to avoid a high point.

plane parallel to the axis of double suction pumps (see Fig. 15.4a). Failure to heed this caution can result in reduced performance, noise, premature impeller erosion, and thrust bearing failure. End suction pumps are more tolerant but are better if the elbow is followed by a reducer, oriented to avoid a high point (see Fig. 15.4b). Two elbows in different planes (Fig. 15.5) produce flow distortion which can take 50 to 100 diameters to dissipate, therefore it is

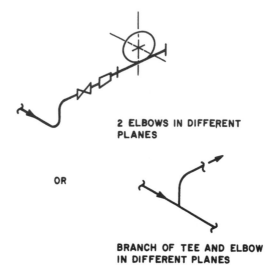

2 ELBOWS IN DIFFERENT
PLANES

OR

BRANCH OF TEE AND ELBOW
IN DIFFERENT PLANES

FIG. 15.5. Undesirable elbow configurations in suction piping.

an arrangement better avoided. Where this is not possible, a downstream flow straightener can offer some improvement, but at the expense of higher friction loss. Suction manifold design needs particular care to avoid connected elbows in different planes. Krutzsch [15.4] addresses the question and provides suggestions for avoiding major problems.

When the liquid is below the pump, it is necessary to provide some means of priming unless the pump is self-priming. And even when the pump is self-priming, it generally has to be provided with some initial liquid to function. Figure 15.6 shows two systems for priming. The simpler arrangement requires a foot valve and a means of filling the suction line and pump, such as a dedicated tank or a line from a separate source or from a common discharge header (provided there are means of priming the first pump). The more complex arrangement is to employ some form of vacuum device, such as an ejector or pump. Such an arrangement is independent of a source of liquid and can be automated when used in conjunction with a liquid level switch. Any pump operating with a suction lift is susceptible to loss of prime. If the pumping service is critical, it is better to have a flooded suction. Where this is not feasible, the pump should be continuously monitored for loss of prime. This can be done with a switch that responds to low pressure, flow, or motor current.

Venting is mandatory if the means of priming involve filling from a separate source or if there is a risk of accumulating gas or vapor at the pump's suction. The latter is a problem with condensate pumps and can be with pumps handling volatile liquids where the suction line is exposed to solar radiation. For pumps handling nonvolatile liquids with a flooded suction, a means of venting (to fill the casing) is advisable if other conditions permit. Vents for filling are valved and piped to either the suction source or a liquid disposal system. Vents for vapor or gas accumulation must be continuous and are therefore piped back to the vapor space of the suction vessel (see Fig. 15.7).

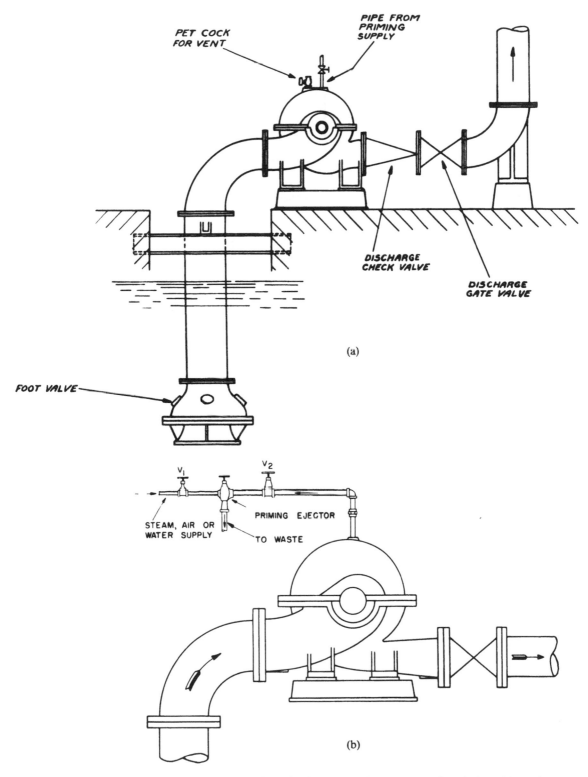

PET COCK
FOR VENT

PIPE FROM
PRIMING
SUPPLY

DISCHARGE
CHECK VALVE

DISCHARGE
GATE VALVE

(a)

FOOT VALVE

V₁

V₂

STEAM, AIR OR
WATER SUPPLY

PRIMING EJECTOR

TO WASTE

(b)

FIG. 15.6.   (a) Installation using foot valve. (b) Arrangement for priming with an ejector.

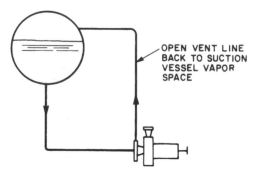

FIG. 15.7.   Suction nozzle vent to prevent vapor or gas accumulation.

As a plant or system is erected, every effort should be made to clean the piping and vessels and to keep them clean. In ideal circumstances the whole system is flushed by circulating liquid with a cleanup pump, having first removed the "elements" of the process pumps. If this is not done, the process pumps become the cleanup pumps, and it is advisable to provide for a temporary suction strainer at each pump. The strainer is usually of conical form (see Fig. 15.8) and must be provided with a means of measuring pressure drop so blockage can be detected.

15.4. Discharge System

The configuration of the discharge system is determined by where the liquid has to be delivered and by the piping loads that can be imposed on the pump. Piping size is determined from a balance between capital cost and pump operating cost; see Kern [in Ref. 1.2] for a detailed treatment. As a general rule, the piping is one size larger than the pump's discharge. Most systems incorporate a check valve to prevent reverse flow and a stop valve to allow the pump to be isolated. When the discharge run is short and there is only one pump, the check valve may be deleted. If isolation is not required, the stop valve also may be deleted.

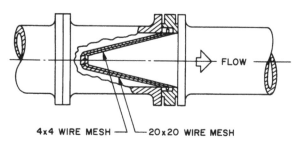

FIG. 15.8.   Typical temporary suction strainer.

15.5. Piping Loads

Unless provided with a flexible connector at each nozzle or resiliently mounted, pumps act as pipe anchors to some degree. A pump's capacity as a pipe anchor can have a profound effect on piping design, hence cost. Piping loads can affect a pump in three ways. First, the nozzles can be overstressed, leading to plastic deformation or fracture. Second, the casing can be distorted to the extent that contact at internal running clearances occurs. Third, the pump's mounting, feet and base, can distort enough to cause severe coupling misalignment. Jones [in Ref. 1.2] provides examples of the third effect. A pump's tolerance of piping loads therefore depends upon its design and construction. Centrifugal pumps with large internal clearances are the most tolerant, the actual degree depending upon the casing material and pump mounting, with close coupled inherently better than separately coupled; see Chapter 6. Although centrifugal pumps are the most tolerant, the loads to which they can be safely subjected are usually well below those acceptable on a pipe stress basis. The notable exception is barrel pumps of up to, say, 10 stages. Rotary pumps have very close internal clearances, thus they are the least tolerant of imposed piping loads. Actual allowable piping loads should be obtained from the pump manufacturer with the balance of the engineering information. When an analysis of piping loads indicates a problem, or the service is one that usually causes a problem, there are three solutions. The first is to flexibly connect the pump to its piping by using expansion joints. The expansion joints must be restrained or else they may impose loads greater than those they were intended to prevent. The second is to change to a more rugged form of pump. In that connection, some purchasers obtain allowable piping loads with the bid data and include that capability in their evaluation. The third is to revise the piping configuration to reduce the loads. The revised configuration is usually more complex, hence more expensive, thereby leading to a balance between piping and pump cost (more rugged pumps also being more expensive).

15.6. Bypass

A bypass may be a means of capacity control; see Chapters 6, 7, and 8. For centrifugal pumps a bypass may also be a means of preventing operation below a flow that will cause rapid pump degradation; see Chapter 6 for a detailed discussion of allowable minimum flow. Figure 15.9 shows a typical bypass flow diagram. Bypass system design has two important aspects. First, each pump must have a separate flow control element. While a common element may seem less expensive, there is every chance that two or more pumps running in parallel would not share the flow (the head characteristic at low flow is subject to wider variations than at BEP and is often flat), resulting in one or more pumps running at zero flow. Second, if the bypass flow is

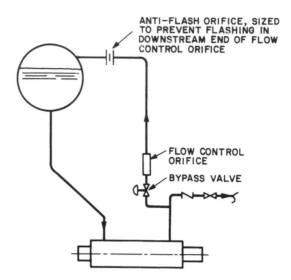

FIG. 15.9.   Typical minimum flow bypass system.

returned to a saturated suction vessel, some means of preventing flashing in the line is necessary. Figure 15.9 shows an antiflash orifice serving this purpose. When a bypass is provided to maintain a minimum flow, then flow should be the control parameter. Pump discharge pressure depends upon the flow through the pump plus the pump's suction pressure, liquid SG, and the pump's head characteristic (which can be flat at low flows), and is therefore not a reliable indicator of pump flow.

15.7. Pressure Relief

Lest the requirement be overlooked, positive displacement pumps must be protected by a FULL CAPACITY relief valve, piped independently of any bypass and without any isolating valve in the line (see Fig. 7.11 and 8.16).

15.8. Warmup

Warmup is justified when there is a risk of pump damage from thermal shock, transient differential thermal expansion, or thermal distortion. Thermal shock is usually not a major problem in typical pump materials below 400°F. At higher temperatures, and at lower temperatures in materials prone

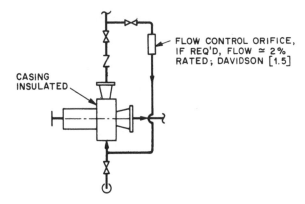

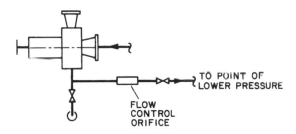

(a) FLOW FROM DISCHARGE HEADER, THRU DRAIN
CONNECTION BACK TO SUCTION.

(b) FLOW FROM SUCTION HEADER, THRU DRAIN CONNECTION
BACK TO POINT OF LOWER PRESSURE.

FIG. 15.10. Common warmup arrangements.

to thermal shock, the pump should be slowly warmed up to operating temperature. Transient differential thermal expansion depends upon the detail design of the pump. Rotary pumps, for example, require warmup whenever the pumping temperature is more than say 50°F above ambient. Centrifugal pumps, having larger clearances, can generally tolerate a sudden temperature rise of 250°F (but consult the manufacturer in case the design is vulnerable). Thermal distortion results when the pump is not at an even temperature throughout. In a standby pump on a service at more than 100°F above ambient, any leakage back into the pump is likely to result in distortion to the extent that there is contact at the internal clearances. Since it is difficult to guarantee there will be no leakage past the discharge check valve, it is better to ensure the pump is completely warmed up. The usual method of warming up is to circulate some process liquid through the pump. Depending upon the system, the liquid can come from the discharge header or the suction vessel (see Fig. 15.10). The quantity circulated and the point(s) of entry into the pump are important; see Heald and Penry [15.5].

15.9. Installing the Pump

As soon as the pump is received, check to see that what has been received is what was ordered. Any discrepancies should be raised immediately with the vendor.

If the pump is going to be stored for some time before installation, it has to be properly protected. Usually this requirement is included in the purchase order. When it's necessary to store a pump that has not been prepared for storage, consult the manufacturer for advice on the procedure to follow.

The installation procedure set out below is of necessity general. It deals with horizontal, separately coupled pumps, since that configuration is the most common and they are among the most difficult to install. Where the equipment differs, use the general procedure as far as it is applicable but be sure to obtain and follow the manufacturer's specific instructions.

Prepare the foundation by roughening the top surface and removing loose material. At each packer or leveling screw location, grind to provide a smooth surface with 70% contact.

Position the base or complete unit using survey marks. When the foundation bolts are not precast in the foundation, they are positioned at this stage with the base. Foundation bolts are better if sleeved as shown in Fig. 15.11. Doing this allows the bolt to be moved slightly and increases its capacity to absorb shock.

Grout the foundation bolts, where applicable, taking care not to grout inside the bolt sleeve.

Level the baseplate. This is one of the most important steps in equipment installation, yet it is frequently overlooked. Failure to level the baseplate can make alignment difficult or impossible to achieve, and the difficulty usually extends beyond installation. Recall from Chapter 13 that bases designed for mounting on a foundation have only moderate bending strength and negligible torsional strength. Leveling is carried out using packers and wedges at each foundation bolt (Fig. 15.12a) or leveling screws acting on packers (Fig. 15.12b). It is easier to level the base with the equipment

FIG. 15.11.   Sleeved foundation bolt (Courtesy Deco).

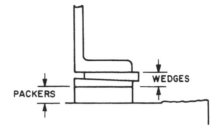

(a) WEDGES

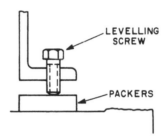

(b) LEVELLING SCREW

FIG. 15.12.   Means of leveling bases.

removed. If removing the equipment poses a problem, leveling can still be done provided the equipment mounting pads extend far enough beyond the feet. Mounting pads should be leveled within 0.0005 in./in., checking in two directions 90° apart; see Monroe [15.6]. Small pumps are sometimes "leveled" by bringing the coupling into alignment. This approach is certainly better than not leveling at all, and it is really all that can be done when the base has unmachined equipment pads.

Grout the baseplate, using either nonshrink cementitious or epoxy grout, prepared and placed in accordance with the manufacturer's instructions. The purpose of grouting is to provide intimate contact between the base and foundation, thus enabling the foundation to support the base as its design intended. See Monroe [15.6] for practical guidance on the placement of epoxy grout. Once the grout has cured, remove the packers or leveling screws and grout the spaces. By doing this, the base is assured of continuous support.

Remount the equipment. When the pump has been factory doweled, use the dowels to accurately reposition the pump, then position the driver, taking care to provide the shaft separation shown on the outline drawing.

Align the equipment. In alignment the objective is to align the axes of rotation, not one shaft or coupling hub to another. To do this *both shafts must be rotated together* lest runouts in the shafts or coupling hubs distort the alignment. Figure 15.13 shows the basic method of determining alignment. There are more sophisticated methods, such as reverse indicators and laser measurement, but the principle is the same. With modern alignment kits the task is made easier by having a preprogrammed computer calculate the equipment movement needed to achieve alignment. The usual practice is to align the driver to the pump, since the pump may not be easily moved once it is connected to its piping. When the outline drawing specifies equipment rise due to

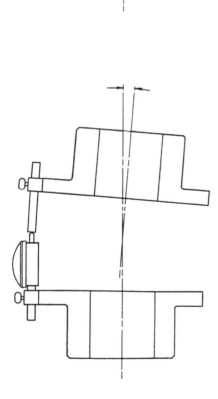

CHECK FOR ANGULAR MISALIGNMENT

DIAL INDICATOR MEASURES MAXIMUM LONGITUDINAL VARIATION IN HUB SPACING THROUGH 360° ROTATION.

1. Attach dial indicator to hub, as with a hose clamp; rotate 360° to locate point of minimum reading on dial; then rotate body or face of indicator so that zero reading lines up with pointer.

2. Rotate both half couplings together 360°. Watch indicator for misalignment reading.

3. Driver and driven units will be lined up when dial indicator reading comes within maximum allowable variation for that coupling style. Refer to specific installation instruction sheet for the coupling being installed. NOTE: If both shafts cannot be rotated together, connect dial indicator to the shaft that is rotated.

CHECK FOR PARALLEL MISALIGNMENT

DIAL INDICATOR MEASURES DISPLACEMENT OF ONE SHAFT CENTER LINE FROM THE OTHER.

4. Reset pointer to zero and repeat operations 1 and 2 when either driven unit or driver is moved during aligning trials.

5. Check for parallel misalignment as shown. Move or shim units so that parallel misalignment is brought within the maximum allowable variations for the coupling style.

6. Rotate couplings several revolutions to make sure no "end-wise creep" in connected shafts is measured.

7. Tighten all locknuts or capscrews.

8. Recheck and tighten all locknuts or capscrews after several hours of operation.

FIG. 15.13. Face and rim dial indicator method of checking coupling alignment (Courtesy Rexnord).

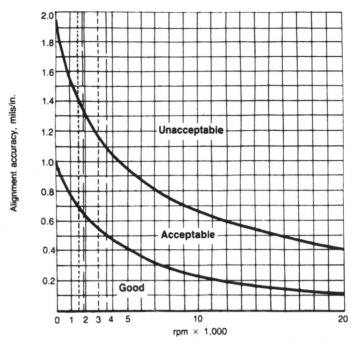

FIG. 15.14. Piotrowski coupling alignment recommendation [15.7].

thermal expansion, the "cold" alignment should allow for this. Allowable misalignment is a function of coupling type, flexible element separation, and rotative speed; see Chapter 12. The values typically quoted by coupling manufacturers tend to be generous and can generally be reduced by a factor of 0.5 and 0.1 without becoming impractical. Bloch [15.7] cites a practical alignment recommendation by Piotrowski (see Fig. 15.14).

Connect the piping. Unless the piping design incorporates "cold spring," the piping should match up to the pump nozzles and those of a turbine driver, if applicable, without strain. In many cases the final piping cut and weld are left until the pump is in place so this can be realized. If prefabricated piping is not matching up, the piping should be corrected by cutting and rewelding rather than by trying to have the pump or turbine withstand additional load.

Check the equipment alignment and make any corrections. If there is major misalignment, the cause must be found and corrected. Typical causes are piping loads, weak baseplates, and poor grouting. If the cause of major misalignment after piping is left uncorrected, it usually manifests itself as continual difficulty with alignment during operation.

Delay doweling the driver until after the pump has been run and the alignment is given a further check. The need to do this explains why there is no justification in shop doweling the driver except in units employing self-supporting bases.

Make up and connect the auxiliary piping: cooling water, seal flush (if from a separate source), lube oil (if from a central system), bypass, and warmup. Follow the manufacturer's outline drawing and piping diagram; be sure to resolve any discrepancies between the drawings and the pump.

Flush the piping systems. The best approach is to remove the critical equipment elements, then circulate liquid through the system. It is important to circulate liquid through the entire system. Blinding off is dangerous; liquid does not circulate through the entire system, which allows debris to accumulate in the "dead" legs. Upon start-up, this debris is washed into the equipment, often with catastrophic results. A further risk with blinding is starting up with the blind still in place, an event that is bound to have catastrophic results.

Inhibit the equipment if it is not going to be put into service immediately. Consult the manufacturer for the appropriate procedure, but at the very least ensure that bearing housings, shaft seals, and the liquid end are inhibited. Do the same for the driver. When the driver is a large electric motor, its stator heaters should be connected and energized.

The last two steps are not strictly "putting the pump in place," but they are fundamental to an orderly start-up. Time constraints often suggest flushing and inhibiting be ignored. Doing so usually winds up costing more in time and lost production than was saved by the "shortcut."

References

15.1. J. R. Brennan, "Controlling Noise in Fluid Pumping Systems," *Plant Eng.*, *28*(4), 89–92 (February 21, 1974).

15.2. Harris, *Handbook of Noise Control*, McGraw-Hill, New York, 1957.

15.3. I. J. Karassik et al., *Centrifugal Boiler Feed Pumps under Transient Operating Conditions* (Paper 53-F-32), ASME Fall Meeting, Rochester, New York, 1953.

15.4. W. C. Krutzsch, "Hydraulic Design Considerations for Pump Suction Piping; Part 1," *Pump World*, *8*(4), 10–15 (1982); "Part 2," *Pump World*, *9*(3), 10–13 (1983).

15.5. C. C. Heald and D. G. Penry, "Design and Operation of Pumps for Hot Standby Service," in *Proc. 5th Pump Symp.*, Houston, May 10–12, 1988, pp. 109–115.

15.6. P. C. Monroe, "Pump Baseplate Installation and Grouting," in *Proc. 5th Pump Symp.*, Houston, May 10–12, 1988, pp. 117–125.

15.7. H. P. Bloch, "Use Laser-Optics for Machinery Alignment," *Hydrocarbon Process.*, pp. 33–36 (October 1987).

16. Operation

As is evident from Chapters 6, 7, and 8, there is a great variety of pumps. All have their operating peculiarities, therefore this section can only address operation in general. The specific instructions given by each manufacturer should be reviewed before any operation, and where contrary to this general treatment, they are to prevail.

16.1 Start-up

Starting a pump involves a certain sequence of steps, depending upon the circumstances of the start. Figure 16.1 shows the usual circumstances and the sequence necessary for each circumstance.

Before operating a pump for the first time on a new service, it is necessary to verify the pump's integrity. Not doing this can result in pump failure during operation, an event that can cause substantial loss of life and property. It is often tempting to "push the button and see what happens," but such a practice is foolhardy and has to be avoided. To verify integrity, check the following:

Are the expected system pressures and temperatures within the pump's pressure containment capability, both casing and seal?

Is the process liquid condition as specified? If not, is the new condition tolerable to the pump?

Is the expected mode of operation, flow, and speed consistent with the pump's rated performance and operating limitations?

With pump integrity verified by either the checks detailed above or by previous operation, make the following detailed equipment checks to determine whether the pump is ready to run:

1. Piping systems cleaned.
2. Inhibiting removed where necessary.
3. Auxiliary piping connected correctly.
4. Coupling alignment, cold, within acceptable limits.
5. Direction of rotation correct. Check motors uncoupled.
6. Equipment lubricated or lubrication system operating.
7. Suction strainer, if installed, clean and equipped with functioning gauges to indicate pressure drop.
8. Pump filled with liquid and all suction valves fully open.
9. Provision for flow from the pump: bypass open, relief valve set correctly and functioning. Small centrifugal pumps may have no provision other than flow to the system. For short-term operation this is acceptable, but the time limit must be allowed for in the start-up cycle.
10. Pump rotates freely. If warmup is provided, this check should be made after the pump has been warmed to uniform temperature.

Once it has been established that the pump is ready to run, start the driver in accordance with the manufacturer's instructions. If the driver is variable speed, bring it up to the minimum operating speed; see specific instructions for the pump. When the pump has been started with its discharge valve shut, open the valve and establish flow to the system. If the system is being filled or charged by the pump, the initial resistance will be low. For positive displace-

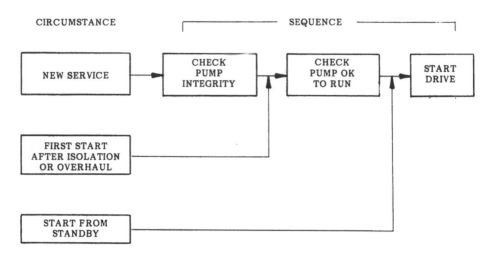

FIG. 16.1. Pump starting qualifications.

ment pumps this does not present a problem. Centrifugal pumps, however, can run out well beyond rated capacity, risking driver overload or pump cavitation. To avoid this, the discharge should be throttled to maintain a minimum discharge pressure until the system is filled. Similar cautions and precautions are applicable to centrifugal pumps arranged for series or parallel operation; see Chapter 6 for details.

Observe pump operation during start-up and make any necessary "settling in" adjustments, such as seal adjustment. Avoid, wherever possible, remote, unattended, start-ups; instrumentation is a poor substitute for being there.

Once operating conditions have stabilized, take and record the following data to establish the pump's baseline behavior:

Suction pressure and temperature
Discharge pressure
Flow (not always easily determined)
Rotative speed (with accuracy)
Driver input, e.g., motor current and voltage, turbine steam conditions
Seal behavior
Bearing or lubricant temperatures
Vibration; housing for antifriction bearings, rotor for sleeve

The purpose of baseline operating data is to provide a reference against which to compare later performance and thereby assess pump condition.

16.2. Running

In routine operation the objective is to conserve pumping energy and equipment. In broad terms, there are three practices that make a substantial contribution to this.

Periodically inspect the pump to ensure it is running normally and to make any necessary minor adjustments. The alternative to periodic inspection is to allow the pump to run until it manifests distress by a gross decline in performance or complete failure. Doing this poses the risk, at the least, of a major interruption in production.

Run only those pumps necessary to maintain the flow. Running more pumps than needed results in higher power consumption and, in many cases, accelerated pump deterioration. Similarly, if one pump proves more than adequate for the flow, seek to have it rerated for a lower capacity or get a smaller pump.

Avoid prolonged operation on the bypass. The energy dissipated across the bypass is wasted, and in the case of centrifugal pumps, prolonged operation at this condition usually results in accelerated deterioration (clearance wear, impeller erosion, bearing fatigue, etc.). If prolonged operation on the bypass is envisaged or becomes necessary, look first to a system change. If that fails, go to a larger capacity bypass; i.e., one sized for continuous operation. The definition of continuous is elusive, but many agree that more than 2 or 3 hours in any 24-hour period constitutes a practical definition.

Various process disturbances can affect pump operation during routine operation. Loss of suction and variations in system resistance are the two most common problems.

Loss of suction can result from quenching a saturated suction vessel, inadequate suction line venting, or running with too low a liquid level in the suction vessel. All these cause a cessation of pumping by way of vapor or gas binding. In pumps dependent upon pumped liquid lubrication of internal clearances, vapor or gas binding also invariably results in severe damage to the clearances or seizure of the pump. Pumps whose clearances don't depend upon pumped liquid lubrication can be allowed to run vapor bound until suction pressure is restored and the vapor is condensed (though at some risk to the shaft seal unless injected from an external source). Gas binding is not so easily corrected: centrifugal pumps must be shut down and vented; reciprocating and rotary pumps may be left running if some liquid is present but they have to be vented.

Variations in system resistance are usually caused by viscosity other than expected, piping friction other than calculated, fouling within process equipment, or strainer blockage. Positive displacement pumps are not unduly affected by such variations; they serve only to vary the pump's discharge pressure and power.

Centrifugal pumps, on the other hand, can suffer substantial flow swings with all their consequences. When process flow is controlled, the risk of running out to high capacities is minimal. But when the system resistance increases, the pump may not develop sufficient head to maintain flow. Under these circumstances the process is slowed and the pump can be forced to run at too low a capacity; see above under operation on the bypass and Chapter 6 for details of the possible consequences of prolonged operation at low capacities.

17. Maintenance

The objective in pump maintenance is conservation of the equipment. By doing this, the equipment yields high availability and high reliability, both at the lowest total cost. There is an unfortunate trend toward cutting down on maintenance expenditure, the motivation being cost reduction. While there may be an initial cost saving, the usual result, except in cases of blatant negligence, is an increase in total cost over the longer term. It is therefore in the general interest to continually highlight the longer term consideration.

In the same manner as operation, and for the same reasons, maintenance can be given only a general treatment in this text. The manufacturer's specific instructions for a particular machine should be referred to and are to prevail.

The essence of good maintenance is attention to detail. Miles [17.1] makes the point and it cannot be overemphasized. Attention to detail starts with establishing and maintaining an accurate and complete equipment record system. The system may be manual or computerized, depending upon plant resources and preference, and there is nothing wrong with a good manual system, but it must be accurate and complete. The value of an equipment record system may not be evident at plant start-up, but it will be once equipment problems develop. Adequate information on equipment history is invaluable for troubleshooting and justifying purchasing decisions that do not reflect the lowest purchase cost. Equipment record systems contain the following information:

Equipment description and purchasing history
Manufacturer's drawing list
"As built" clearances and materials
Start-up data
Preventive maintenance history
Overhaul history, including any changes made

17.1. Preventive Maintenance

The procedures covered by this term are carried out periodically to keep the machine running at its optimum. In order of decreasing frequency, the usual procedures for pumps are:

Seal adjustment: packed box and adjustable Chevron seals need adjustment to take up wear and minimize leakage. Not adjusting such seals results in high leakage and can lead to major damage to seals or bearings or both if allowed to persist.

Lubrication: bearings and "mechanisms" require lubrication. In some arrangements the assembly has to be dismantled, cleaned, and relubricated, usually with grease. Other arrangements use oil, which has to be changed. The frequency of lubrication depends upon the machine, the service, and the environment in which it is operating.

Wear: some pump types, centrifugal pumps with axial clearances at the impeller for example, can be adjusted to compensate for clearance wear. This procedure would be employed whenever the pump showed an obvious drop in performance, or perhaps periodically until a trend was determined.

Drive: coupling alignment tends to change with time, so periodic checking and correction is warranted. V belt drives require retensioning to compensate for stretch. Failure to attend to the drive can result in serious damage to both the drive and the connected equipment.

The preventive procedures given above do not cover drivers; see the manufacturer's instructions for such procedures.

17.2. Overhaul

The age-old question concerning overhauls is When? There are three approaches. One is to dismantle the machine periodically, regardless of condition, and replace any parts that look dubious. The argument for this practice is that it essentially assures the machine will run trouble-free until the next overhaul. Provided the examination of the parts is sufficiently thorough and the period consistent with expected component life, this approach is valid. An example of it is aircraft maintenance. When examination of the parts is not so thorough and the period is chosen randomly, periodic overhaul can result in a false sense of security. Component failure by metal fatigue is the usual shortcoming.

The second approach is to "run it till it breaks." When the service is not critical or the pump is fully spared (and the spare is immediately available), and the added cost of consequent damage in the breakdown can be tolerated, such an approach can be lived with. There are, however, a lot of "ifs," and it is questionable whether the total cost is really as low as claimed. Compounding that, the cavalier element inherent in this approach often carries over into the overhaul process, leading to breakdowns more frequently than ought to be the case.

A refinement of the second approach yields the third. By seeking to learn as much about the machine's condition while it is still running, it is possible to develop a reasonably accurate profile of its condition and, using a little judgment, to determine when it should be taken out of service for overhaul. The data used to develop this profile are performance and behavior, hence the importance of establishing these data at pump start-up. Behavior, i.e., vibration, temperature, and noise, are useful data, but are often lagging indicators of trouble. When behavior is combined with performance, the picture is more complete and usually clearer.

TABLE 17.1   Guide to Pump Damage

Liquid Condition	Damage
Clean	Normal: clearances eroded, minimal damage to other hydraulic parts, shaft seal, and bearings
	Abrasive erosion: solids in liquid
	Corrosion erosion: mild corrosion
	Cavitation erosion: insufficient NPSH or operation at low flows
	Adhesive erosion (scoring at clearances): high rotor forces, pump distortion, running dry
	Short seal life: unsuitable seal, poor seal environment or seal angularity
	Short bearing life: low flow operation, operation at over pressure (positive displacement), lubricant contamination from high seal leakage or atmosphere, coupling misalignment, insufficient cooling, worn clearances, undersize bearings
Corrosive	Normal: clearances eroded, minimal metal loss from other parts unless pump intentionally sacrificial or being used at limit of material
	Abnormal: see above for clean liquid
Abrasive	Normal: abrasive erosion of all hydraulic parts, severity depending upon service conditions, seal similar
	Abnormal: similar to clean liquid after allowing for abrasive service, need to distinguish between damage from abrasives and other causes.

Of the three approaches, the third is the most favored, since it promises tolerable availability and reliability at reasonable expense.

Patience is fundamental to successful overhaul. There are other more spectacular approaches, but the spectacle tends to wear off with repetition.

Before the pump is removed, check and record its coupling alignment and shaft spacing. Disconnect major piping with care and note any major strain. Take the pump or element to a clean shop, and dismantle it in an orderly sequence.

Once the pump is dismantled, examine the parts carefully. Visual inspection is good for wear, at least qualitatively. Measure critical parts, recording the dimensions on a form suitable for the pump type, and determine the "worn" running clearances. Inspect dynamic parts by NDE such as liquid penetrant or magnetic particle. Do the same with pressure containment parts if there are any suspicious results from the visual inspection. Look for and note signs of abrasive or cavitation erosion.

Compare the pump's condition with what could be considered "normal" for the service. This is difficult task because many factors influence pump wear. There are, however, reasonable classifications of "normal," and it is worth the trouble to try to get a better indication of whether what has been encountered is to expected or not. The "not" is the important part because that says there is scope for improvement, which may well constitute a cost saving for the plant. Finding the root cause of an abnormal problem will require quite a deal of dedication, but it is worth the effort just to avoid having to needlessly overhaul a pump time and time again.

In a very general sense, Table 17.1 can serve as a guide to "normal" pump damage and the likely causes of abnormal damage. The guide is based on liquid condition. See Bloch [17.2] for a detailed treatment.

Correct, wherever possible, the cause of abnormal damage, In many instances the cause is relatively simple, requiring only perseverance to find it. At the same time, inquire into progress if "normal" damage is proving hard to tolerate. Since the pump was built, there may have been progress in a direction that would help limit the damage.

Returning a pump to new condition is usually done by a combination of replacing and restoring parts. Some parts, badly damaged impellers for example, cannot be easily restored and are therefore routinely replaced. Other parts, such as gear pump side plates, can usually be faced off to restore the surface to essentially new condition. When there is a choice between restoring and replacing, the following factors need to be considered:

Cost of replacement versus restoration
Availability of replacement parts
Integrity of restored parts
Effect of restoration on performance

In some instances the integrity of restored parts can be superior to that of replacements. Such cases usually reflect a "short cut" in the original equipment purchase, and could probably be rectified if the manufacturer were made aware of the field problem being encountered.

Rebuild the pump to original specifications unless there are very sound and well researched reasons to do otherwise. The term "well researched" should include consultation with the manufacturer. On many occasions, users of equipment have made well engineered changes and realized a notable improvement in service life. There are, unfortunately, just as many instances where exactly the opposite has been the case. If there is a rule to be followed in refining machine design, it is: "no one has a monopoly on ideas."

Reinstall the pump, taking the same care with piping strain and coupling alignment as required for the original installation. When whole pumps are exchanged and by definition not returned to their original position, piping strain can be a problem. The problem arises because pump casings have, of necessity, a tolerance in their dimensions. When nozzle locations are determined by castings, the tolerances can be quite large. Pumps that are overhauled by restoring or exchanging an "element" (assembly made up of the rotor plus stationary parts) avoid this problem.

Put the overhauled pump back into service using the start-up sequence set out in Chapter 16, Operation, for such circumstances.

Lest it be overlooked, add the overhaul data and new baseline conditions to the equipment record system.

References

17.1.   D. Miles, "Pump Maintenance: Paying Attention to Details," *Plant Serv.*, pp. 18–20, 22 (September 1987).

17.2.   H. P. Bloch, and F. K. Geitner, *Machinery Component Maintenance and Repair*, Gulf Publishing Company, Houston, Texas 1985.

D	diameter (in.)
g	gravitational constant (ft/sec^2)
H	head (ft)
P	pressure or power (lb/in.^{2} [psi]) (BHP)
Q	flow rate (US gal/min)
SG	specific gravity (ratio)
V	velocity (ft/sec)
η	efficiency (ratio)
ρ	density (slug/ft^3)

Subscripts

a	atmosphere for pressure
	acceleration for head
g	guage
G	guage
L	loss (due friction)
s	static
t	total
v	vapor
z	elevation

Milton Keynes UK
Ingram Content Group UK Ltd.
UKHW052018071024
449327UK00027B/2331

9 780367 403140